CLIMATE CRISIS AND THE
GLOBAL GREEN NEW DEAL

CLIMATE CRISIS AND THE GLOBAL GREEN NEW DEAL

The Political Economy of Saving the Planet

Noam Chomsky and Robert Pollin,
with C. J. Polychroniou

VERSO
London • New York

First published by Verso 2020
© Noam Chomsky, Robert Pollin,
C. J. Polychroniou 2020

1 3 5 7 9 10 8 6 4 2

Verso
UK: 6 Meard Street, London W1F 0EG
US: 20 Jay Street, Suite 1010, Brooklyn, NY 11201
versobooks.com

Verso is the imprint of New Left Books

ISBN-13: 978-1-78873-985-6
ISBN-13: 978-1-78873-987-0 (UK EBK)
ISBN-13: 978-1-78873-986-3 (US EBK)

British Library Cataloguing in Publication Data
A catalogue record for this book is available from the British Library

Library of Congress Cataloging-in-Publication Data

Names: Chomsky, Noam, author. | Pollin, Robert, author.
Title: The climate crisis and the global green new deal : the political
  economy of saving the planet / Noam Chomsky and Robert Pollin with
  Chronis Polychroniou.
Description: First edition paperback. | London ; New York : Verso Books,
  2020. | Includes bibliographical references and index. | Summary: "An
  inquiry into how to build the political force to make a global Green New
  Deal a reality"— Provided by publisher.
Identifiers: LCCN 2020021851 (print) | LCCN 2020021852 (ebook) | ISBN
  9781788739856 (paperback) | ISBN 9781788739863 (ebk)
Subjects: LCSH: Green movement—Political aspects. | Green
  movement—Economic aspects. | Climate change mitigation—Political
  aspects. | Climate change mitigation—Economic aspects. |
  Capitalism—Environmental aspects.
Classification: LCC JA75.8 .C46 2020 (print) | LCC JA75.8 (ebook) | DDC
  363.738/7461—dc2 3
LC record available at https://lccn.loc.gov/2020021851
LC ebook record available at https://lccn.loc.gov/2020021852

Typeset in Adobe Garamond Pro by Hewer Text UK Ltd, Edinburgh
Printed and bound by CPI Group (UK) Ltd, Croydon CR0 4YY

Contents

Introduction

Since the origins of civilized social order, the human race has faced a full gamut of severe challenges and deadly threats, ranging from famines and natural disasters (floods, earthquakes, volcanic eruptions, and so on) to enslavement and wars. In the first half of the twentieth century, humanity experienced two world wars and the emergence of the greatest genocidal regime ever. Over the second half of the twentieth century, we have lived with the threat of nuclear annihilation hanging over our heads like Damocles' sword. As I write in April 2020, we face the global COVID-19 pandemic and accompanying economic collapse. Nobody knows at this point how many people will die as a result of the pandemic. We also cannot yet know how severe will be the subsequent recession. The signs point to a crisis of at least the severity of the 2007–09 Great Recession and perhaps comparable to the 1930s Depression.

Nonetheless, a strong case can be made that humanity faces its greatest existential crisis ever with climate change. That is, trapped carbon dioxide and other greenhouse gases resulting, first and foremost, from burning oil, coal, and natural gas to generate energy, are raising average temperatures in all regions of the globe. The consequences of a hotter planet include increasing incidences of heat extremes, heavy precipitation, droughts, sea level increases, biodiversity losses, and corresponding impacts on health, livelihoods, food security, water supply, and human security. Meanwhile, climate denialism maintains a strong grip over much of the human race, especially in the United States. This is due in part to the fossil fuel industry's relentless propaganda and obfuscation campaigns over decades. It is also linked to the unlikely outcome of Donald Trump, the Climate-Denier-in-Chief, somehow making it into the White House with his November 2016 election victory over Hillary Clinton. President Trump has gone so far as to declare global warming a "hoax" and to pull the United States out of the 2015 Paris Climate Agreement, which was endorsed by 195 countries, including the United States under Barack Obama.

Still, one cannot deny the impact that fear of the unknown and the potential loss of jobs may be exerting on people when they resist the reality of global warming. This is exactly why it is so important that any plan to effectively combat the climate crisis must include provisions that ensure workers are able to make a fair transition to a carbon-free economy. More specifically, any version of the widely discussed Green New Deal project must include these priorities:

1. Greenhouse gas emissions reductions will at least achieve the targets set in 2018 by the Intergovernmental Panel on Climate Change, namely a 45 percent reduction in global emissions by 2030 and the attainment of net zero emissions by 2050.

2. Investments to dramatically raise energy efficiency standards and equally dramatically expand the supply of solar, wind, and other clean renewable energy sources will form the leading edge of the transition to a green economy in all regions of the world.

3. The green economy transition will not expose workers in the fossil fuel industry and other vulnerable groups to the plague of joblessness and the anxieties of economic insecurity.

4. Economic growth must proceed along a sustainable and egalitarian path, such that climate stabilization is unified with the equally important goals of expanding job opportunities and raising mass living standards for working people and the poor throughout the world.

A global Green New Deal that includes these four priorities is, in fact, the only viable solution available to us if we hope to avoid the catastrophic repercussions of persistently rising average global temperatures. Given the absence of such a coherent Green New Deal program, all international climate summits that have occurred thus far, including the twenty-fifth UN-sponsored Conference of the Parties (COP25) held in Madrid in December 2019, have failed to put the world onto a viable climate stabilization path. Even the much-celebrated COP21 conference in Paris in 2015 mainly produced another round of ritual inaction. Because of these failures, the world is already hotter by over 1 degree Celsius (1.8 degrees Fahrenheit) above preindustrial levels, and on its way to 1.5°C (2.7°F) warmer within another decade or two.

The catastrophic consequences that will result from unchecked climate change are described in detail in the analyses found in this book by its two authors, Noam Chomsky and Robert Pollin. Noam Chomsky, of course, has been the world's leading public intellectual for more than half a century now. He is also the father of modern linguistics. His work in that field has exerted tremendous influence in a wide variety of other fields, including mathematics, philosophy, psychology, and computer science. Robert Pollin is a world-renowned progressive economist who has been a leader fighting on behalf of an egalitarian green economy for more than a decade. He has produced a large number of important publications as well as commissioned studies on implementing Green New Deal programs in countries around the world as well as multiple US states. He also served as a consultant to the US Energy Department on implementing the green investment components of the 2009 American Recovery and Reinvestment Act, the Obama economic stimulus program that included $90 billion in funding for investments in renewable energy and energy efficiency.

The global Green New Deal program that Pollin outlines in this book is strongly endorsed by Chomsky. Pollin shows how all four criteria listed above for such a program are readily achievable, when considered strictly in terms of the technical and economic obstacles to be overcome. Beyond all such technical and economic challenges, the most daunting obstacle to success is mounting the necessary political will to defeat the gigantic vested interests and resources of the global fossil fuel industry.

This book includes four chapters. Chapter 1, titled "The Nature of Climate Change," begins by situating the challenge of global warming alongside other crises that the human race has faced in the past. The chapter then offers detailed critiques on an array of major questions, such as why market-driven proposals to tackle the climate crisis are doomed to failure and why alternatives to industrial agriculture are of immense importance to reaching a viable climate stabilization path.

Chapter 2, titled "Capitalism and the Climate Crisis," presents clear theoretical and empirical discussions of the connections between capitalism, environmental destruction, and the climate crisis. It also offers valuable insights on whether capitalists' werewolf hunger for profits can be, in any way, reconciled with the imperative of stabilizing the climate. This chapter also examines the reasons why political action has thus far failed to make significant advances in tackling the crisis.

Chapter 3, titled "A Global Green New Deal," describes the program that is needed to achieve a successful transition to a green economy. Pollin sketches out what a global Green New Deal entails and how it can be financed. He also describes the ways through which such a program can become a bulwark against the long-term rise of inequality that has prevailed under forty years of global neoliberalism. Pollin also provides a critical assessment of the European

Union's own plan for what it has termed its "European Green Deal." Chomsky then closes the chapter by considering the nightmarish scenario of millions of people from the global South trying to migrate to the high-income countries of the global North as the catastrophic effects of global warming intensify with time.

The fourth and last chapter of the book is titled "Political Mobilization to Save the Planet." It addresses questions such as how the climate crisis might affect the global balance of power, whether eco-socialism has the potential as a politico-ideological vision to mobilize people in the struggle to create a green future, and what the connections are between climate change and the 2020 COVID-19 pandemic. The overarching question that animates this chapter is the most basic one: what needs to be done to advance a successful political mobilization on behalf of a global Green New Deal.

In my view, this little book that the reader holds in their hands is critically important. There should be food for thought in it for everyone—scholars, activists, and lay people alike. Of course, it is only one modest contribution toward a public dialogue that must continue to expand until it reaches all levels of society in all regions of the globe. Pushing that global dialogue forward, even by a small amount, is the least we all owe to the next generations. With that in mind, I wish to extend my most heartfelt thanks and deepest gratitude to Noam Chomsky and Robert Pollin for allowing me to travel with them on this journey to help inform the public about how we can all save the planet.

C. J. Polychroniou
April 2020

The Nature of Climate Change

Over the last couple of decades, the challenge of climate change has emerged as perhaps the most serious existential crisis facing humanity but, at the same time, as the most difficult public issue for governments worldwide. Noam, given what we know so far about the science of climate change, how would you summarize the climate change crisis vis-à-vis other crises that humanity has faced in the past?[1]

Noam Chomsky: We cannot overlook the fact that humans today are facing awesome problems that are radically unlike any that have arisen before in human history. They have to answer the question whether organized human society can survive in any recognizable form. And the answers cannot be long delayed.

The tasks ahead are indeed new, and dire. History is all too rich in records of horrendous wars, indescribable torture, massacres, and every imaginable abuse of fundamental rights. But the threat of destruction of organized human life in any recognizable or tolerable form—that is entirely new. It can only be overcome by common efforts of the entire world, though of course responsibility is proportional to capacity, and elementary moral principles demand that a special responsibility falls on those who have been primarily responsible for creating the crises over centuries, enriching themselves while creating a grim fate for humanity.

These issues arose dramatically on August 6, 1945. Though the Hiroshima bomb itself, despite its horrendous effects, did not threaten human survival, it was apparent that the genie was out of the bottle and that technological develop-ments would soon reach that stage—as they did, in 1953, with the explosion of thermonuclear weapons. That led to the setting of the Doomsday Clock by the *Bulletin of Atomic Scientists* at two minutes to midnight—meaning global termination—a dread setting to which it returned after Trump's first year in office, describing the next year as "the new abnormal."[2] Prematurely. In January 2020, thanks largely to Trump's leadership, the clock was moved closer to midnight than ever before: 100 seconds, dropping minutes for seconds. I won't run through the grim record, but anyone who does will recognize that it is a near miracle that we have survived thus far, and the race to self-destruction is now accelerating.

There have been efforts to avert the worst, with some success, notably four major arms control treaties: ABM, INF, Open Skies, and New START. The Bush II administration withdrew from the ABM Treaty in 2002. The Trump admin-istration withdrew from the INF Treaty in August 2019, timing its withdrawal almost exactly with Hiroshima Day. It has also indicated that it will not maintain the Open Skies or New START Treaties.[3] That will mean that all bars are down and we can race toward terminal war.

The general "reasoning"—if one can use that word for total madness—is illustrated by the US withdrawal from the INF Treaty, followed predictably by Russia's own withdrawal. This major treaty was negotiated by Reagan and Gorbachev in 1987, greatly reducing the threat of war in Europe, which would quickly become global, hence terminal. The US claims that Russia is violating the treaty, as the media regularly report—failing, however, to add that Russia claims that the US is violating the treaty, a claim taken seriously enough by US scientists that the authoritative *Bulletin of Atomic Scientists* devoted a major article to expounding it.[4]

In a sane world, the two sides would move to diplomacy, bringing in outside experts to evaluate the claims, and then reaching a settlement, as Reagan and Gorbachev did in 1987. In an insane world the treaty would be abrogated and both sides would merrily proceed to develop new and even more dangerous and destabilizing weapons, such as hypersonic missiles, against which there is no currently imaginable defense (if there are defenses against any major weapons systems, a dubious prospect).

Our world.

Like the INF Treaty, the Open Skies Treaty was a Republican initiative. The idea was proposed by President Eisenhower, and implemented by President George H. W. Bush (Bush I). That was the pre-Gingrich Republican Party, still a sane political organization. Two respected political analysts of the American Enterprise Institute, Thomas Mann and Norman Ornstein, describe the Republican Party since Newt Gingrich's takeover in the '90s as not a normal political party but a "radical insurgency" that has largely abandoned parliamentary politics.[5] Under Mitch McConnell's leadership, that has only become more evident—but he has ample company in Party circles.

The abrogation of the INF Treaty elicited little reaction apart from in arms control circles. But not everyone is looking the other way. The military industry can scarcely conceal its delight over the huge new contracts to develop means to destroy everything, and the more far-sighted are also developing longer-term plans to gain fat contracts to develop possible (if unlikely) means of defense against the monstrosities they are now free to develop.

The Trump administration wasted no time in flaunting its abrogation of the treaty. Within a few weeks, the Pentagon tersely announced the successful launching of an intermediate range missile violating the INF Treaty—virtually inviting others to join in, with all of the obvious consequences.[6]

Former Defense Secretary William Perry, who has spent much of his career on nuclear issues and is not given to exaggerated rhetoric, declared some time ago that he was "terrified," in fact doubly terrified—both by the increasing threat of war and the slight attention it receives. We should in fact be triply terrified, adding that the race to terminal destruction is being carried out by people who are fully aware of the horrendous consequences of what they are doing. Much the same is true of their dedicated efforts to destroy the environment that can sustain life.

The net spreads wide. It is not just the policy makers, the Trump administration being particularly egregious and dangerous. It reaches to the big banks that are pouring money into fossil fuel extraction, and the editors of the best journals running article after article about the wondrous new technology that has propelled the US to the lead in producing the substances which will destroy us unless radically curbed, all without mention of the terrible word "climate."

Scientists seeking extraterrestrial intelligence have been struck by the Fermi paradox: Where are they? Astrophysics suggests that there should be intelligent life elsewhere. Maybe they are right; there really is intelligent life, and when it discovers the strange inhabitants of Planet Earth, it has the sense to stay far away.

Let's keep however to the second major threat to survival, environmental catastrophe.

It was not understood at the time, but the early post–World War II period marked a turning point in a second threat to survival. Geologists generally take the early post–World War II period to be the onset of the Anthropocene, a new geological epoch in which human activity is having a profound, and devastating, impact on the environment, a judgment on timing confirmed most recently in May 2019 by the working group on the Anthropocene.[7] By now evidence of the severity and imminence of the threat is overwhelming—and is quietly recognized even by the most extreme deniers, as we see below.

How are the two existential crises related? A simple answer is given by Australian climate scientist Andrew Glikson: "Climate scientists are no longer alone in having to cope with the global emergency, whose implications have reached the defence establishment, yet the world continues to spend near to $1.8 trillion each year on the military, a resource that needs to be diverted to the protection of life on Earth. As the portents for major conflicts—in the China Sea, Ukraine, and the Middle East are rising—who will defend the Earth?"[8]

Who indeed.

Climate scientists are certainly paying close attention and issuing frank and explicit warnings. Oxford professor of physics Raymond Pierrehumbert, a lead author of the frightening 2018 Intergovernmental Panel on Climate Change (IPCC) report (since superseded by more urgent warnings), opens his review of existing circumstances and options by writing: "Let's get this on the table right away, without mincing words. With regard to the climate crisis, yes, it's time to panic . . . We are in deep trouble." He then lays out the details carefully and scrupulously, reviewing the possible technical fixes and their very serious problems, concluding that "there's no plan B."[9] We must move to zero net carbon emissions, and fast.

The deep concerns of climate scientists are readily available to those who don't prefer to hide their heads in the sand. CNN celebrated Thanksgiving 2019 with a detailed (and accurate) report of an important study that had just appeared in *Nature* on tipping points—moments at which the dire effects of global warming will become irreversible. The authors conclude that consideration of tipping points and their interactions reveals that "we are in a climate emergency and strengthens this year's chorus of calls for urgent climate action . . . The risk and urgency of the situation are acute . . . The stability and resilience of our planet is in peril. International action—not just words—must reflect this."[10]

The authors warn further that "atmospheric CO_2 is already at levels last seen around four million years ago, in the Pliocene epoch. It is rapidly heading towards levels last seen some 50 million years ago—in the Eocene—when temperatures were up to 14°C higher than they were in pre-industrial times." And what happened over very long periods then is now compressed by human action to a few years. They explain further that existing forecasts, while grim enough, have failed to take into account the effects of tipping points.

They conclude that "the intervention time left to prevent tipping could already have shrunk towards zero, whereas the reaction time to achieve net zero emissions is 30 years at best. Hence we might already have lost control of whether tipping happens. A saving grace is that the rate at which damage accumulates from tipping—and hence the risk posed—could still be under our control to some extent."

To some extent, and there is no time to lose.

Meanwhile the world watches as we proceed toward a catastrophe of unimaginable proportions. We are approaching perilously close to the global temperatures of 120,000 years ago, when sea levels were six to nine meters higher than today.[11] Truly unimaginable prospects, even discounting the effect of more frequent and violent storms, which will put paid to whatever wreckage is left.

One of many ominous developments that might fill the gap between 120,000 years ago and today is the melting of the vast West Antarctic ice sheet. Glaciers are sliding into the sea five times faster than in the 1990s, with more than 100 meters of ice thickness lost in some areas due to ocean warming, and those losses doubling every decade. Complete loss of the West Antarctic ice sheet would raise sea levels by about five meters, drowning coastal cities, and with utterly devastating effects elsewhere—the low-lying plains of Bangladesh, for example.[12]

Only one of the many concerns of those who are paying attention to what is happening before our eyes.

Dire warnings from climate scientists abound. Israeli climatologist Baruch Rinkevich captures the general mood succinctly: "After us, the deluge, as the saying goes. People don't fully understand what we're talking about here . . . They don't understand that everything is expected to change: the air we breathe, the food we eat, the water we drink, the landscapes we see, the oceans, the seasons, the daily routine, the quality of life. Our children will have to adapt or become extinct . . . That's not for me. I'm happy I won't be here."[13]

Rinkevich and his Israeli colleagues discuss various likely "horror scenarios" for Israel, but a few are optimistic. One observes that "Israel is definitely not the Maldives and it is not expected to be submerged anytime soon." Good news. They generally agree, however, that the region may become mostly unlivable: "Cities are liable to be abandoned in Iran, Iraq and in developing countries, but in our country it will be possible to live." And although the temperature of the Mediterranean may approach 40°C (i.e., 104°F), "the maximum permitted temperature in a Jacuzzi," nevertheless, "humans will not be boiled alive like sea urchins and red-mouthed rock shells, but there could be mortal danger during the height of the bathing season."

So there is hope for Israel under the most optimistic forecasts, if not for the region.

The essential observation is made by Professor Alon Tal: "We are aggravating the condition of the planet. The Jewish state has looked humanity's ultimate challenge in the eyes and said: 'Forget it.' What will we tell our children? That we wanted a higher quality of living? That we had to remove all the natural gas from the sea because it was so economically profitable? Those are pathetic explanations. We're talking about the most fateful issue there is, especially in the Mediterranean Basin, and the government of Israel isn't capable of appointing a minister who cares that we are simply going to be cooked."[14]

Tal's comment is correct, and deeply troubling. What is it about humans that makes them able to accept "pathetic explanations" or just say "forget it" while looking "humanity's ultimate challenge in the eyes"? That's the response whether it is gradual impending environmental catastrophe or the opportunity to construct new means to destroy us all at once. What is it about humans that enables them to spend $1.8 trillion on the military—the US far in the lead—while not asking, "Who will defend the earth?"

While Tal's observation generalizes, it is somewhat too strong. There are countries, and localities, where serious efforts are being undertaken to act before it is too late. And it is not too late. The answer to the mad race to produce more means of self-destruction is obvious enough, at least in words; implementation is another matter. And there is still time to mitigate the impending climate catastrophe if a firm commitment is undertaken. That is surely not impossible if the facts can be faced. In 1941, the US faced a serious though incomparably lesser threat, and responded with a voluntary mass mobilization so overwhelming that it greatly impressed Nazi Germany's economic czar Albert Speer, who lamented that totalitarian Germany could not match the voluntary subordination to the national task in the more free societies.

Some estimate that the challenge, while immense, does not impose burdens comparable to those of 1941. Economist Jeffrey Sachs, in a careful study, concludes that "contrary to some commentaries, decarbonization will not require a grand mobilization of the U.S. economy on par with World War II. The incremental costs of decarbonization above our normal energy costs will amount to 1 to 2 percent of U.S. GDP per year during the period to 2050. By contrast, during World War II, federal outlays soared to 43 percent of GDP from the prewar level of 10 percent of GDP in 1940."[15]

It can be done, but now we face a cruel irony of history. Just at the time when all must act together, with dedication, to confront humanity's "ultimate challenge," the leaders of the most powerful state in human history, in full awareness of what they are doing, are dedicating themselves with passion to radical escalation of the twin threats to survival. The government is in the hands of the only major "conservative party in the world that rejects the need to tackle climate change" and is also opening the door to the development of new and more threatening weapons of mass destruction.[16]

The members of the astonishing troika who have the fate of the world in their hands are the Secretary of State, the National Security Adviser, and the Chief—from the perspective of the world, the Godfather; international relations resemble the Mafia to an extent rarely recognized. The Secretary of State, Mike Pompeo, is an Evangelical Christian whose acuity as a political analyst is revealed by his belief that God may have sent Trump to the world to save Israel from Iran.[17]

The National Security Adviser until his September 2019 resignation (or firing, depended on whom you choose to believe), was John Bolton, who has left his minions in place. Bolton had a simple doctrine: the US must accept no external limits on its freedom of action—no treaties, no international agreements or conventions—and therefore must ensure that every country will have maximal opportunity to develop the means to destroy us all—the US in the lead, for what that's worth. He also flaunts a corollary: bomb Iran because it will never agree to negotiate on anything.[18] This prescription for action, and predication, was confidently issued while Iran was negotiating with the US and Europe on the Joint Comprehensive Plan of Action (JCPOA), the detailed agreement finalized shortly after freezing Iranian nuclear activities—an agreement observed meticulously by Iran, as US intelligence and others confirm, and torn up by the Chief.

The Chief is an infantile megalomaniac, and very effective con man, who couldn't care less if the world burns or explodes, as long as he can pretend to be the winner as he two-steps over the cliff waving his little red hat triumphantly.

Trump's reasoning on the environment was well expressed after he was prevented from building a golf course with luxurious homes because it would have endangered the drinking water supply for nearby communities. As he explained to an appreciative crowd of realtors, "I was building a development. I was going to build some really luxury, beautiful houses. [But] I found out that I can't build on the land. Does that make sense to you?" What then could be more reasonable than to roll back dozens of environmental regulations, significantly increasing greenhouse emissions, including "the nation's benchmark [Nixon-era] environmental law," which would mean that federal agencies "no longer have to take climate change into account when they assess the environmental impacts of highways, pipelines and other major infrastructure projects"? And, by extension, what could be more reasonable than to maximize the use of fossil fuels with the understanding that it will soon undermine the prospects for organized human life on Earth?[19]

And they are not alone in the world scene. In what may be a symbolic inauguration, the new year 2020 opened with the continent of Australia on fire and people desperately fleeing temperatures of a blast furnace during record-breaking heat, while the prime minister—a dedicated denialist—returned grudgingly from a vacation to assure his constituents that he felt their pain. Meanwhile the opposition labor leader toured the coal plants, calling for expansion of Australia's role as world champion coal exporter and assuring the country that this was quite consistent with Australia's serious commitment to combating global warming—a commitment that, according to international monitors, places it last among fifty-seven countries on climate change policy.[20]

We may contemplate how history could have conjured up such a nightmare, but here we are.

Trump has good reason to revel in his success—whatever the cost to the irrelevant population of the world. His primary constituency—great wealth and corporate power—may not like him, but they are quite happy with the gifts that he lavishes on them. And his voting base is mesmerized. Over half of Republicans regard Trump as the greatest of US presidents, surpassing Lincoln, the former champion.[21] The impeachment proceedings seem to have improved his standing among the faithful, supporting the thesis that dark forces are seeking to undermine their leader—who many of them believe has come (or maybe has been sent) to rescue them from the neoliberal assault that he in reality champions vigorously. An impressive conjuring trick.

These are people who are going to have to be convinced of the urgency of the threats we face if there is to be any hope of escaping disaster.

It may seem rather oxymoronic to try to conjure up a geopolitical strategy emanating from this cast of characters —I'll skip the "advise and consent" Senate, where the Republican majority, having largely abandoned any lingering integrity, is firmly in Trump's pocket, terrified of angering his fervent base. But a strategy does emerge from the clouds: the construction of a reactionary international, run from the White House, bringing together the vicious military and family dictatorships of Egypt and the Gulf; Israel consummating its Greater Israel project now with the open rather than tacit support of the US; Modi's India, crushing Kashmir and dismantling what remains of India's secular democracy in favor of an extremist Hindu nationalist ethnocracy; Bolsonaro's Brazil, with a stream of ugly crimes but none approaching his commitment to destroy the Amazon, "the lungs of the earth," by handing it over to his friends in agribusiness and mining; and a slew of other attractive members, like Orbán's Hungary, celebrating its nomadic Magyar roots back to Attila the Hun if not Genghis Khan, and Salvini's Italy, righteously murdering thousands of miserable people fleeing from Libya, famous as the site of Italy's genocidal exploits under Mussolini.[22] And who knows what might be in the wings—conceivably Farage taking over the US vassal state that was once Britain if Boris Johnson's hard Brexit proceeds on what seems a likely course.

While this might be the shape of the emerging world, like the environmental crisis it is not at all inevitable. There are choices, and they can make a huge difference.

One choice was announced by Bernie Sanders and Yanis Varoufakis, the former finance minister of Greece under the leftist Syriza government, who together issued a call for a progressive international to confront the reactionary international being forged under the Trumpian aegis. It should not be allowed to fall on deaf ears.

Returning to the original question, the brief answer is that the environmental crisis, along with its twin nuclear crisis, is unique in human history, and is a true existential crisis. Those alive today will decide the fate of humanity—and the fate of the other species that we are now destroying at a rate not seen for 65 million years, when a huge asteroid hit the earth, ending the age of the dinosaurs and opening the way for some small mammals to evolve to become finally the asteroid's clone, differing from its predecessor in that it can make a choice.

Bob, IPCC produced a special report in 2018 on the impact of global warming of 1.5ºC above preindustrial levels. In your view, do mainstream studies on the challenges of climate change, as those undertaken over the years by the IPCC, capture adequately the nature and the risks of climate crisis?

Robert Pollin: Of course, I am not a climate scientist, and therefore I am not qualified to assess the mainstream work that is regularly summarized in IPCC studies, as opposed to studies that believe that the IPCC is not adequately representing where the science is at any given moment. But let's understand what role the IPCC plays in the world of advancing and disseminating climate science research. The IPCC is a UN agency created in 1998 to fulfill its stated mission "to provide policymakers with regular scientific assessments on the current state of knowledge about climate change."[23] The IPCC does not carry out original research but rather serves as a clearinghouse for assessing and synthesizing the relevant literature. Thousands of scientists contribute to writing and reviewing the IPCC's reports, which are then reviewed by governments. I myself know well the climate scientists at my own university, University of Massachusetts Amherst, who are involved in various IPCC projects. These are very committed, capable, and credible people. So it is fair to say that the IPCC does bring together current, high-quality assessments of mainstream climate science on any given set of questions.

There remains a small band of climate deniers, whose positions are given credence and then amplified in the mainstream media far beyond what is warranted given the scientific findings they have produced.[24] Nevertheless, while it is implausible, we cannot totally rule out the possibility that some of their positions may have merit. But, exactly to this point, it is also the case that the IPCC is scrupulous in recognizing a high degree of uncertainty in all of its estimates. For example, its targets for the needed level of emissions reductions are never presented as a single figure, as in, say, "we must reduce emissions by 80 percent within twenty years or face these certain terrible consequences." Rather, the IPCC

always presents its conclusions in terms of ranges and probabilities. It is also true that the IPCC has regularly changed
its assessments to a significant degree, as illustrated in recent
years by some of its most important publications.

Thus, in its 2007 Fourth Assessment Report, the IPCC
concluded that in order to stabilize the global average (mean)
temperature at 2.0°C above the preindustrial average, annual
CO_2 emissions needed to fall to, roughly, between 4 and 13
billion metric tons by 2050—that is a decline of 60–88
percent relative to the 2018 level of 33 billion metric tons.
However, in its Fifth Assessment Report, released in 2014,
the IPCC set the range of the necessary emissions reduction
at 36–76 percent to achieve the same 2.0°C stabilization
point. In other words, between the IPCC's 2007 and 2014
Assessment Reports, its emissions reduction requirement had
declined. However, in 2018, four years after the 2014
Assessment Report, the IPCC dramatically shifted its position again, taking a much more alarmist stance than in its
previous publications. That is, in the October 2018 report to
which you refer, titled *Global Warming of 1.5°C*, the IPCC
emphasized the imperative of limiting the increase in the
global mean temperature to 1.5 rather than 2.0 degrees. It
did so after having reached the conclusion that hitting the 1.5
degree target will dramatically lower the likely negative consequences of climate change. These include the risks of heat
extremes, heavy precipitation, droughts, sea level rise, biodiversity losses, and the corresponding impacts on health, livelihoods, food security, water supply, and human security.

It is clear that there is a high degree of uncertainty as to
the full set of consequences we face by allowing average
global temperatures to rise above 1.5 degrees or even 2.0

degrees above preindustrial levels. The consequences could be more severe or less severe than those predicted by the IPCC's 2018 assessment. If it were to change its assessment yet again in future reports, that would be true to form. So the IPCC could conceivably offer more sanguine appraisals, as in 2014, but it is more likely that it will become still more dire in its appraisals, reflecting the warning of the eminent climate scientist and IPCC author Raymond Pierrehumbert, cited above by Noam, that "yes, it's time to panic. We are in deep trouble." In short, we have more than enough information to take decisive action now on the basis of what we know, in full recognition of the range of uncertainties we face.

To pursue that question a bit further, wouldn't it make sense if we applied the insurance option to the problem of tackling climate change?

RP: The short answer is yes, absolutely. Dealing with the reality of uncertainty on these matters does raise the question: What if the overwhelming consensus of scientific opinion turns out to be wrong, or, more precisely, that the relatively low-probability outcome that there will be no serious consequences resulting from climate change turns out to be the actual outcome? Will the global community then have effectively wasted trillions of dollars over a thirty-year period to solve a problem that never existed in the first place?

In fact, we need to take decisive action now on climate change, not based on 100 percent certainty as to its consequences, but rather through estimating reasonable probabilities. Indeed, we should think of a global Green New Deal as exactly the equivalent of an insurance policy to protect ourselves and the planet against the serious prospect—though not the certainty—that we are facing an ecological catastrophe.

The late Harvard economist Martin Weitzman, who died in 2019, contributed important research on how we should handle the uncertainties surrounding climate change. As he put it in *Climate Shock*, his 2015 book with coauthor Gernot Wagner, "climate change is beset with deep-seated uncertainties on top of deep-seated uncertainties on top of still more deep-seated uncertainties." Weitzman and Wagner offer this analogy on how to handle such uncertainties:

> If a civilization-as-we-know-it altering asteroid were hurtling toward Earth, scheduled to hit a decade hence, and it had, say, a 5 percent chance of striking the planet, we would surely pull out all the stops to try to deflect its path. If we knew that same asteroid were hurtling toward Earth a century hence, we may spend a few more years arguing about the precise course of action, but here's what we wouldn't do: We wouldn't say that we should be able to solve the problem in at most a decade, so we can just sit back and relax for another 90 years. Nor would we try to bank on the fact that technologies will be that much better in 90 years, so that we can probably do nothing for 91 or 92 years and we'll still be fine. We'd act, and soon. Never mind that technologies will be getting better in the next 90 years, and never mind that we may find out more about the asteroid's precise path over the next 90 years

that may be able to tell us that the chance of hitting Earth is "only" 4 percent rather than the 5 percent we had assumed all along.[25]

Weitzman and Wagner also bring the issue down to the everyday situations people now face in dealing with uncertainty and insurance, writing that "devastating home fires, car crashes, and other personal catastrophes are almost always much less likely than 10 percent. And still, people take out insurance to cover against these remote possibilities, or are even required to do so by laws that hope to avoid pushing these costs onto society."[26]

From this perspective, the only major issue in dispute with respect to purchasing climate insurance is how much we should be willing to pay to carry sufficient coverage. This is the equivalent of deciding not *whether* to purchase automobile insurance but, rather, how much to spend and how much coverage we need. This is the question I will take up later in the context of describing a viable Green New Deal project.

The underlying premise of orthodox economics is that the operations of free markets, left to their own devices, will produce social outcomes that are superior to government interventions. To what extent is this pro-market bias in orthodox economics holding back progress in climate change mitigation?

RP: In 2007, Nicholas Stern, the prominent mainstream British economist and former chief economist at the World Bank, wrote that "climate change is a result of the greatest market failure the world has seen." Stern's assessment was extreme, but not hyperbolic.

Neoliberalism is a driving force causing the climate crisis. This is because neoliberalism is a variant of classical liberalism, and classical liberalism builds from the idea that everyone should be granted maximum freedom to pursue their self-interest within capitalist market settings. But neoliberalism also diverges substantially from classical liberalism, and therefore also from the basic premises of orthodox economics that free markets, left to their own devices, will produce outcomes that are superior to government interventions. Here is the problem with neoliberalism, when counterposed against a purely free market model celebrated by economic orthodoxy. That is, what really occurs in practice under neoliberalism is that governments allow giant corporations to freely pursue profit opportunities to the maximum extent. But then government fixers arrive on the scene to bail out the corporations whenever their profits might be threatened. This amounts to socialism for capitalists, and harsh, free market capitalism for everyone else.

The oil companies' record in dealing with climate change represents a dramatic case study of neoliberalism in practice. In 1982, researchers working at the then Exxon Corporation (now ExxonMobil) estimated that by about 2060, burning oil, coal, and natural gas to produce energy would elevate the planet's average temperatures by about 2°C. This, in turn, would generate exactly the types of massive climate disruptions that we have increasingly experienced since the 1980s. In 1988, researchers at Shell Corporation reached similar conclusions. We now know what Exxon and Shell did with this information—they buried it. They did so for the obvious reason that, if the information were then known, it might have threatened their prospects for receiving massive profits from producing and selling oil.

There is no minimizing the fact that what Exxon and Shell did was immoral. But it is equally clear that both companies behaved exactly according to the precepts of neoliberalism—that is, they acted to protect their profits. They also continued from the 1980s onward to behave according to the precepts of neoliberalism in extracting the largest possible subsidies that they could get from any and all governments throughout the world. Amid all this, neither company has faced any government sanctions for its behavior. Quite the contrary, they have continued to earn huge profits and receive huge government subsidies.

Now we can't blame this all on orthodox economics. As Stern emphasized, it is also possible, within orthodox economics, to recognize that market processes under capitalism can fail. But it is critical to point out here that orthodox economists insist we address market failures through minimizing the extent of government interventions since, in their view, on balance government interventions are more likely to make things worse, through incompetence or corruption or, still more fundamentally, a fuzzy goal of trying to improve social welfare. By contrast, with markets, nobody is kidding anybody—we are all just out to get the most for ourselves.

This is why virtually all mainstream economists support a carbon tax as the most effective, and for many of them, the *only* effective, policy intervention for fighting climate change. Thus, a January 2019 statement signed by twenty-seven Nobel Prize–winning economists, along with four former chairs of the Federal Reserve and fifteen former chairs of the president's Council of Economic Advisers, asserted:

> A sufficiently robust and gradually rising carbon tax will replace the need for various carbon regulations that are less efficient. Substituting a price signal for cumbersome regulations will promote economic growth and provide the regulatory certainty companies need for long-term investment in clean-energy alternatives.[27]

These economists do agree to redistribute the revenue from the carbon tax back to the population in equal shares, thereby preventing the tax from raising the cost of living for lower-income people, who spend a significant share of their overall income on purchasing energy. However, these economists offer no support for increases in public investments in renewable energy and energy efficiency, thereby surrendering the power of the public sector, amounting to 35 percent of GDP in the US and higher shares elsewhere, to push the clean energy transformation forward at the most aggressive possible rate. They also oppose regulations that require electrical utilities to stop burning coal and natural gas and expand their renewable energy capacities. These positions amount to massive policy errors, committed by many of the most prestigious economists in the United States.

Speaking of massive policy errors committed by prestigious economists, we need to give special prominence to William Nordhaus of Yale, who received the Nobel Prize in Economics in 2018 for his decades of highly influential research, on precisely the economics of climate change. In his Nobel Prize lecture, delivered in December of that year,

Nordhaus presented alternative policy scenarios for addressing climate change. On what he terms in the lecture the "optimal" policy path, the global average temperature will rise by 2°C as of 2050, but then continue to rise for the next one hundred years, reaching the "optimal" stabilization point of a 4°C average global temperature increase by 2150. In other words, to begin with, Nordhaus gives no credence to the conclusion published by the IPCC in October 2018—just two months before his Nobel lecture—that we need to set the average temperature stabilization target at 1.5°C, not 2°, by 2050 to avoid facing intensifying risks with respect to heat extremes, heavy precipitation, droughts, sea level rise, and biodiversity losses. But still more alarming—or let's say, breathtakingly shocking—is that Nordhaus is utterly sanguine about accepting the risks we would face by allowing the global mean temperature to rise by 4°C by 2150.

Surveying the body of research on what the world could plausibly look like with 4°C of warming, the science journalist Mark Lynas writes:

> At four degrees another tipping point is almost certain to be crossed . . . This moment comes as the hundreds of billions of tons of carbon locked up in Arctic permafrost—particularly in Siberia—enter the melt zone, releasing globally warming methane and carbon dioxide in immense quantities . . . The whole Arctic Ocean cap will also disappear, leaving the North Pole as open water for the first time in at least three million years. Extinction for polar bears and other ice-dependent species will now be a certainty. The south polar cap may also be badly affected . . . This would eventually add another five meters

to global sea levels . . . As the sea level rise accelerates, coastlines will be in a constant state of flux. Whole areas, indeed whole island nations, will be submerged. In Europe, new deserts will be spreading in Italy, Spain, Greece, and Turkey: the Sahara will have effectively leapt the Straits of Gibraltar.[28]

It is true, as Martin Weitzman's work on climate uncertainty emphasizes, we have no way of knowing for sure how likely it is that these outcomes would result through allowing the global average temperature to rise by 4°C. But, still following Weitzman, or better still, just applying basic common sense, we should know enough to realize that we must take every action to prevent 4°C of warming to occur, even as a low-probability event. The fact that the single most prominent orthodox economist in the world working on climate change considers the risks from 4°C of warming to be "optimal" tells us everything about the bankrupt state of orthodox economics.

There is growing concern regarding the impact of industrial agriculture on the environment. In fact, the system of industrialized food seems to be bad for human health and the economy overall. Bob, what are the impacts associated with industrial agriculture, and what's the alternative?

RP: Corporate industrial agriculture is a major driver of climate change, responsible for roughly 25 percent of total greenhouse gas emissions, including CO_2, methane, and nitrous oxide, the three main greenhouse gases.[29]

But before getting into some details on climate change issues per se, I need to at least mention some additional major impacts of industrial corporate agriculture. As described in an excellent recent study by the International Labour Organization, industrial agricultural has become a major contributor to

> soil degradation (the loss of organic matter as a result of overexploitation and mismanagement), desertification and freshwater scarcity (through inadequate land and crop management), biodiversity loss, pest resistance and water pollution (resulting from change in land use, eutrophication [i.e., over-enrichment of water with minerals and nutrients, which induces excessive growth of algae], run-off and improper nutrient management).[30]

These sources of soil degradation and water pollution in turn contribute to a range of human health problems. Most critically, hundreds of millions of agricultural workers worldwide are now exposed on a daily basis, and at close quarters, to toxic pesticides and herbicides. From there, toxic substances flow into the food and water supply consumed by the general population.

Returning to the climate impacts of industrial agriculture, there are four major interrelated channels to emphasize: (1) deforestation; (2) the use of land for cattle farming, which consumes far more of the available earth's surface than any other purpose, including growing crops for food; (3) the heavy reliance on natural-gas-based nitrogen fertilizers along with synthetic pesticides and herbicides to increase land productivity; and (4) the huge amount of food that is grown but wasted. Massive food wastage occurs in both low- and high-income countries, though for mainly distinct reasons.

Deforestation

RP: Other than the burning of fossil fuels to produce energy, deforestation is the most significant force driving climate change, owing to the fact that living trees absorb and store CO_2. When trees are felled through deforestation, the CO_2 that is stored in them is released into the atmosphere. In addition, of course, trees that have been chopped down are no longer available to absorb CO_2. As of the most recent 2019 data reported by the IPCC, these combined effects of deforestation—the releasing of CO_2 into the atmosphere from felled trees and the loss of those trees as CO_2 absorbers—are themselves responsible for about 12 percent of all greenhouse gas emissions.

Given that deforestation is fully understood as a major cause of climate change, we need to ask why the practice continues. The answer is straightforward; indeed, it's no more complicated than understanding why we keep burning fossil fuels even though we know it pushes us toward ecological catastrophe. That is, there are profits to be made from destroying forests, because it creates major open land areas that can be exploited for agriculture and mining.

The single biggest profit opportunity created by deforestation is to clear the land for corporate farming. A recent detailed study by Noriko Hosonuma and coauthors estimates that about 40 percent of all deforestation in developing countries is driven by corporate agricultural imperatives, with the most important of these to open land for cattle grazing. Growing cash crops like palm oil for the global market offers another big profit opportunity. Hosonuma estimates that another 33 percent of deforested land is used for subsistence farming. Another 10 percent of the deforested land is used to build roads and other infrastructure, which are, of course, mainly needed to support the new business activities in the deforested areas. Thus, about 85 percent of deforestation is tied to agriculture, mostly cattle farming and other forms of corporate agriculture.[31]

We need to recognize that deforestation can raise incomes for low-income people and communities, in addition to delivering profits for corporations. But these benefits for working people and the poor are almost entirely short-run effects that evaporate quickly. The gains and losses to low-income people from deforestation follow a familiar boom-and-bust cycle. At first, the newly cleared lands attract investments in farm and mining projects as well as the infrastructure needed to support these new businesses. These investments generate jobs, but only so long as the initial phases of getting the projects up and running continue. But even during these initial phases of project development and construction on the newly cleared land, what also happens is that more people migrate to these areas looking for work. This creates increased competition for the newly available jobs and downward pressure on wages.

Any possible benefits to low-income people from expanding subsistence farming and employment in corporate farms are also counteracted by the loss of income-generating activities provided by the forest itself. These include tapping trees for rubber, nut farming, and gathering lumber on a sustainable basis from dead trees. A 2018 study by the World Resources Institute thus concludes that "land acquisition for commodity production often displaces local livelihoods without respect for indigenous and traditional land rights."[32]

The major policy initiative to stop deforestation is a family of globally scoped policies collectively known as REDD—Reducing Emissions from Deforestation and Forest Degradation. This combination of policies is mostly administered and coordinated by the United Nations (UN-REDD) and the World Bank. The basic idea behind REDD is simple: to reward governments, companies, forest owners, and forest dwellers in the global South for keeping their forests intact instead of cutting them down.

In principle, REDD programs should be beneficial. But major problems have emerged in practice. I will just mention three of the most significant ones. First, REDD projects are substantially funded by corporations looking for carbon offsets. For example, REDD can allow an electric utility to purchase carbon credits, which then enables it to continue to burn coal to generate electricity, as opposed to directly transitioning out of coal to investments in high-efficiency and clean, renewable energy sources. Related to this is the second problem, that of "leakages." This refers to REDD initiatives that establish a given set of forests as out of bounds for land clearance, which then prompts businesses to move their operations to other locations that are unprotected. Estimates of leakage rates linked to specific protected sites vary widely—from a negligible amount to over 100 percent of avoided emissions. At the least, it is clear that safeguards to prevent leakages remain weak.[33]

A third major problem with REDD programs is that the financial rewards for participating in them flow disproportionately to corporate farmers and land speculators who understand the legal hoops they must jump through to obtain the available benefits. The actual forest dwellers do not typically have access to the legal and financial advisors who could help them work the system to their benefit. Fair and effective policies for stopping and reversing deforestation are, of course, an imperative. But it is also critical that REDD programs not be rigged to serve the same class of corporate interests who have been benefiting from clearing forests and burning fossil fuels in the first place.

Cattle Farming

RP: Cattle farming contributes to climate change via two channels. The first results from the fact that cattle farming requires far more land than any other form of agriculture. That is, producing food from all other animal sources such as chicken, pork, and fish, as well as growing crops intended directly for human consumption rather than cattle feed, all require far much less land than raising cattle. Cattle farming can be a net positive contributor to the world's overall food supply when the cattle graze only on pastures in which crops cannot grow. But massive amounts of the earth's total land resources are wasted when areas suitable for cultivating food crops for people are instead devoted to either cattle grazing or growing animal feed. Creating this pressure to devote more land to cattle grazing in turn incentivizes corporations and land speculators to clear the forests.

In addition to creating these pressures on land use, raising cattle itself contributes to climate change because cows emit methane gas through their normal digestive processes. This is true for all ruminant animals, that is, animals that regurgitate food and re-chew it, including sheep, goats, buffalo, deer, elk, giraffes, and camels. But the global population of cows and bulls is about 1.5 billion, far greater than the other ruminants. The cows are responsible for about 2 billion tons of greenhouse gas per year through their methane emissions. This alone amounts to about 4 percent of total greenhouse gas emissions as of 2018.

Industrial vs. Organic Farming

RP: Conventional industrial agriculture methods depend on the heavy use of synthetic fertilizer, irrigation, pesticides, and herbicides. The use of nitrogen fertilizer alone increased by 800 percent between 1961 and 2019. Over this same sixty-year period, this practice has been a major contributing factor to the 30 percent increase in the per capita global food supply.

But it is also the case that the manufacturing of nitrogen fertilizer, mainly in the form of ammonia, relies on mixing the hydrogen in natural gas with the nitrogen in the air. As such, manufacturing nitrogen fertilizer produces CO_2, methane, and nitrous oxide, the three main greenhouse gases. In addition, nitrogen fertilizer converts to nitrous oxide when it is combined with soil bacteria.

As an alternative to these industrial agricultural practices, organic farming relies on crop rotation, animal manures and composting for fertilizer, and biological pest control. More specifically, legumes are planted to fix nitrogen in the soil, as opposed to relying on ammonia for nitrogen enhancement, natural insect predators are encouraged as opposed to synthetic pesticides, crops are rotated to confuse pests and .

renew soil, and natural materials are used to control diseases and weeds. The carbon footprint of organic farming is minimal because it does not rely on using ammonia for fertilizer or other fossil-fuel-derived products.

The benefits of organic farming with respect to emissions reduction and climate change are therefore straightforward. But organic farming as an alternative to conventional practices does also present problems that we cannot gloss over. The most critical of these is that food productivity for a given area of land is generally lower than conventional agriculture. How much lower is a matter of dispute. Several large-scale studies have been conducted to answer this question. The range of estimates does vary. Among other factors, the relative productivity differences will depend on regions of the world and the circumstances specific to any given farm. Nevertheless, as a general conclusion, a reasonable midpoint of these various estimates would be that conventional methods produce between about 10 and 15 percent more food for a given area of agricultural land. Yet it is also the case that some researchers find that organic farming is more productive than conventional farming in developing countries, because the materials needed for organic farming are more accessible than synthetic materials in many poor countries.

Wasting Land and Food

RP: As a general point, it is reasonable to assume that producing the world's food supply through organic methods is likely to require more land. This then also reinforces the need to transition away from cattle farming as the single most prevalent use of agricultural land worldwide.

If the world is going to transition into organic farming over industrial agriculture, another important way to counteract any consequent pressures on land use is to significantly reduce the amount of food that is grown but wasted. According to estimates, between about 35 and 50 percent of total global food produced is discarded, degraded, or consumed by pests instead of being eaten. Developing countries typically lose more than 40 percent of food post-harvest or during processing because of inadequate storage and transport infrastructure. High-income countries don't experience this extent of food wasted at the level of production. But it is still estimated that more than 40 percent of food supplied in high-income countries is wasted at the level of retail distribution and consumption. As cases in point, lots of food goes uneaten in restaurants and shoved down the garbage disposal in people's homes.

For developing countries, a first obvious solution to this problem is to improve storage and transportation infrastructure. Even reducing wastage in developing countries by, say, 10 percent will itself reduce the demand for all global land supply by around 5 percent. This alone would compensate significantly for any requirements for additional land use in moving from industrial to organic farming as the primary source of the world's food supply. Within high-income countries, to simply stop wasting so much food that has already been prepared in restaurants and private kitchens would have a comparable major impact in reducing global pressures on land use.

I need to raise one final issue. That is: Should people change their diets as one component of a global Green New Deal, and, in particular, significantly reduce their beef consumption? Inescapably, the answer is "yes." Shifting away from beef to chicken, pork, or fish as animal products, or more ambitiously to a mainly vegetarian diet, will lead to a corresponding decline in the demand for land to support cattle farming. This will reduce incentives for deforestation. It will also mean lower methane gas emissions from cows, since the global cattle population will shrink along with the demand for beef products. Finally, it will diminish any advantages corporate industrial agriculture currently holds over organic farming in terms of the amount of food produced within a given area of land.

In addition to being the primary cause of climate change, burning fossil fuels to produce energy is also the major cause of air pollution. How severe has the problem of air pollution become in terms of health impacts globally?

RP: Air pollution is a severe health hazard throughout the world. According to a 2019 study by the Health Effects Institute, more than 90 percent of the world's population breathes unsafe air, as measured by the World Health Organization's Air Quality Guideline.[34] It is therefore not surprising that, after high blood pressure, smoking, and high blood sugar, air pollution is the fourth leading risk factor for death, responsible for roughly 5 million deaths worldwide in 2017. For low-income countries, air pollution is the number one risk factor for death. It is also the fourth leading cause globally of "disease burden"—that is, the number of years people live in poor health—with roughly 150 million people experiencing premature morbidity or disability due to air pollution.

Air pollution can be divided into two distinct categories—outdoor and indoor pollution. Burning oil, natural gas, and especially coal to produce energy is the most significant cause of outdoor air pollution, in addition to being the main cause of climate change. Coal combustion releases toxic levels of sulfur dioxide and soot particulates, while burning oil and natural gas as well as coal releases toxic amounts of nitrogen oxide into the atmosphere. Another major source of outdoor pollution is wildfires. Climate change, in turn, causes more frequent and severe wildfires, such as those that burned dramatically in Northern California in 2019, and with still greater force in Australia in 2020. These intense wildfires result from the combination of weather extremes—heavier-than-normal rainy periods producing excessive vegetation that then becomes dry fuel during extended periods of heat waves and drought. A third main factor contributing to outdoor air pollution is the burning of biomass to produce energy. Overall, then, the project to build a global clean energy infrastructure to supplant both fossil fuels and high-emissions bioenergy will also serve to eliminate most major sources of outdoor air pollution.

Indoor air pollution is also caused by burning biomass fuel sources—firewood, crop waste, and dung—for cooking and heating. This is done almost exclusively in poor rural households in low-income countries. Indoor air pollution is therefore not as directly linked as outdoor pollution to the burning of fossil fuels. Nevertheless, a clean energy transformation, in which cheap electricity is delivered to rural areas through small-scale solar or wind energy installations, will also eliminate the need for households to burn biomass in their homes. This will, in turn, also eliminate indoor air pollution.

The health risks from air pollution have declined significantly on a global basis for the past thirty years. Thus, in 1990, the number of deaths due to air pollution was 111 per 100,000 people. By 2017, this figure had fallen to 64 deaths per 100,000 people. However, almost all of this improvement has resulted from reductions in indoor air pollution—that is, fewer households burning solid fuel for cooling and heating. There has been almost no improvement in reducing health risks from outdoor air pollution over this same period. Indeed, absent a transition to a global clean energy infrastructure, the impacts of outdoor air pollution are likely to worsen over time. This is because of the rising proportion of the overall population in low-income countries migrating from rural areas into cities. The cities experience worsening air pollution as part of their current economic growth trajectory, dominated by fossil fuel energy sources.

A report written by University of Massachusetts Amherst economist James Boyce focusing on the specific situation in New Delhi in 2015 vividly captures conditions throughout the main urban centers in rapidly growing low-income countries:

> One of the most dangerous air pollutants is particulate matter. In Delhi it comes from multiple sources, including diesel trucks that are allowed to pass through the city in the middle of the night, rapidly growing numbers of passenger vehicles, coal-burning power plants and brick kilns that ring the city, construction debris and open burning of wastes. Particulates are measured by an Air Quality Index (AQI). An AQI below 50 is considered "good." Anything above 300 is considered "hazardous" and would trigger emergency warnings in many countries. In Delhi I

soon fell into the habit of checking data [on AQI] from our nearest location. When I checked on the morning of Valentine's Day, the AQI for particulates was 399. Overnight it had hit 668, off the standard AQI chart. Sometimes it soared even higher. A month before I arrived in Delhi, the Centre for Science and Environment, India's leading environmental advocacy organization, released the results of a study in which several residents were equipped with handheld devices to monitor air pollution levels as they went about their activities in the course of a typical day. Some of the readings topped 1,000.[35]

For the most part, conditions are certainly less severe in high-income countries. Thus, the average death rate due to air pollution in the US and Germany is one-seventh that of India, while Japan is still lower at one-twelfth the Indian death rate. Nevertheless, the health hazards from outdoor air pollution are still substantial in most high-income countries. They also vary widely within high-income countries by both class and race. Here as well, Boyce along with his coauthors have done pioneering work in documenting these disparities, focusing on those in the US. In a 2014 study, they found that, in the US Midwest for example, poor people of color face roughly twice the exposure to toxic air than non-poor whites. Poor whites face about 13 percent greater exposure than non-poor whites. But non-poor people of color face roughly 30 percent greater exposure than even poor whites.[36]

Overall, then, it is abundantly clear that air pollution and climate change are deeply interconnected. Stabilizing the climate through a global Green New Deal can also solve most air pollution problems and the severe health problems that accompany air pollution. The benefits of eliminating most forms of air pollution will also flow disproportionately to low- and middle-income countries as well as to low-income people and minorities in high-income countries. This then is one of the ways in which the global Green New Deal can become a unified program for both human equality and ecological sanity.

2

Capitalism and the Climate Crisis

The 2015 UN climate agreement signed in Paris, popularly known as the COP21 agreement, has been hailed by world leaders (with the exception of Donald Trump) as a huge diplomatic success, but it has been rightly criticized by environmentalists and others for lacking any teeth. There's indeed nothing mandatory in the Paris Agreement. Noam, why is it so difficult to check climate change?

NC: Looking beyond COP21, there is a great deal to say about why checking climate change is so difficult. But as to why the limited Paris Agreement has no teeth, the answer is clear enough.

The original goal was to establish a treaty with binding commitments. Laurent Fabius, the summit's president, reiterated that goal strongly. But there was a barrier: the US Republican Party, which controlled Congress, would not accept any meaningful arrangements.

The Republican leadership was admirably frank about its intention to undermine the Paris Agreement. One reason, hardly concealed, is that the Republican wrecking ball must smash anything done by the hated Obama; that goal was put plainly by Senate Majority Leader Mitch McConnell when Obama was elected. Another reason is the principled opposition to any external constraints on US power. But the immediate decision follows directly from the Party leadership's uniform rejection of any efforts to confront the looming environmental crisis—a stand traceable in large part to the historic service of the Party to private wealth and corporate power, accelerated during the neoliberal years.

Following carefully worked-out Republican plans, McConnell informed foreign embassies that, according to sources reported in *Politico*, "Republicans intend to fight Obama's climate agenda at every turn." He also made it clear that any agreement that reached the GOP-controlled Senate would be "dead on arrival." "There is 'no chance' that such an agreement could clear the two-thirds hurdle, one Republican energy lobbyist said. 'There are few certainties in life, but that is one of them.' " Republicans also made it clear that they would "block Obama's pledge to provide billions of dollars to help poor countries adapt to the effects of a warming planet," and would sabotage other efforts to deal with global warming. "They are becoming the party of climate super-villains," as one commentator put it succinctly.[1]

It's important to recognize the nature of this organization. For anyone who did not yet understand, it was made crystal clear during the 2016 Republican primaries, featuring political figures who were hailed as the cream of the crop—apart from the interloper who walked off with the prize to the dismay of the Republican establishment. Every single candidate either denied that what is happening is happening, or said maybe it is but it doesn't matter (the latter message came from the "moderates," former governor Jeb Bush and Ohio governor John Kasich). Kasich was considered the most serious and sober of the candidates. He did break ranks by recognizing the basic facts, but added that "we are going to burn [coal] in Ohio and we are not going to apologize for it."[2]

That's 100 percent support for destroying the prospects for organized human life, with the most respected figure taking the most grotesque stand.

Amazingly, this astonishing spectacle passed with virtually no comment (if any) within the mainstream, a fact of no little import in itself.

It's of some interest to see how this remarkable situation came to pass. There are general reasons (there is no space to go into those here), but also quite specific—and revealing—ones. A decade ago the Republican organization, while already well off the normal spectrum of parliamentary politics, was not firmly dedicated to denying what the leadership knows to be true. How this changed provides some insight into contemporary politics, under the impact of the most dedicated and reactionary elements of the highly class-conscious business world.

A glimpse into this world was provided after the death of David Koch in August 2019. This happened to coincide with the appearance of a major in-depth study of the Koch empire and corporate America by Christopher Leonard, who discussed some of his discoveries in articles and interviews.

Leonard describes David Koch as the "ultimate denier," whose rejection of anthropogenic global warming was deep and sincere. Let us put aside suspicions that this might have something to do with the fact that he had an immense fortune at stake in this denialism, perhaps trillions of dollars of potential losses over a period of thirty years or more if denialism were to fail, Leonard estimates.[3] Let's nevertheless suspend disbelief and accept that the convictions were entirely sincere. That would come as little surprise. John C. Calhoun, the grand ideologist of slavery, was no doubt sincere in believing that the vicious slave labor camps of the South were the necessary foundation for a higher civilization. And there are other examples, which, out of politeness, I will not mention.

The Koch brothers' denialism went vastly beyond mere efforts to convince. They launched huge campaigns to ensure that nothing would be done to impede the exploitation of the fossil fuels on which their fortune rests. As Leonard recounts, "David Koch worked tirelessly, over decades, to jettison from office any moderate Republicans who proposed to regulate greenhouse gases."[4] But the efforts did not entirely succeed. In 2009–10, the Republicans were flirting with reality, coming close to supporting a market-based cap-and-trade plan for greenhouse gas emissions. John McCain ran for president on the Republican ticket in 2008 warning about climate change. With the help of Mike Pence and others like him, the Koch juggernaut was able to derail the heresy, ridding the Party of moderates who might not toe the line on denialism and twisting the arms of the recalcitrant with a combination of public vilification and private funding. The consequences of which we now see before us. The lessons about "really existing democracy" as well.

The Koch network, Leonard writes, "has tried to build a Republican Party in its image: one that not only refuses to consider action on climate change but continues to deny that the problem is real." With impressive success.

The juggernaut is indeed impressive. No stone was left unturned: networks of rich donors, discourse-shifting think tanks, one of the largest lobbying groups in the country, the organization of what can pose as grassroots groups to ring doorbells, pretty much creating and shaping the Tea Party. And it had many other goals as well, such as undermining labor rights, destroying unions, and blocking government policies that might help people: what's called "libertarianism" in US usage.[5]

The Koch brothers' juggernaut stands out in its careful planning and successful use of the immense profits it has gained from polluting the global atmosphere without cost—a mere "externality," in the terminology of the trade. But it is symbolic of the savage capitalism that is becoming more and more evident as the neoliberal project that has served private wealth and corporate power so well comes under threat.

Both political parties have drifted right during the neoliberal years, much as in Europe. The Democratic establishment is now more or less what would have been called "moderate Republicans" some years ago. The Republicans have mostly gone off the spectrum. Comparative studies show that they rank alongside fringe right-wing parties in Europe in their general positions. They are, furthermore, the only major conservative party to reject anthropogenic climate change, as already mentioned: a global anomaly.

The positions of the leadership on climate surely influence the attitudes of Party loyalists. Only about 25 percent of

Republicans (36 percent of the savvier millennials) recognize that humans are responsible for global warming.[6] Shocking figures. And in the ranking of high-priority issues among Republicans, global warming (if it is even assumed to be taking place), remains low and unchanged into the election year.

It might be considered outrageous to assert that today's Republican Party is the most dangerous organization in human history. Perhaps so, but in the light of the stakes, what else can one rationally conclude?

Even apart from Republican obstructionism, it is unlikely that the US would have accepted binding commitments at Paris. The US rarely ratifies international conventions, and when it does, it is typically with reservations that exclude the US. That's true even of the Genocide Convention, signed by the US after forty years but excluding the US, which retains the right to commit genocide. There are many other examples.

Returning to COP21, the immediate reason for the lack of teeth is the Republican Party, but the chances that the US would have agreed to binding commitments were slim even without the obstructionism of the most dangerous organization in world history.

In the background of this obstructionism is a lingering question, the one raised by Alon Tal: Why is it so hard for governments to confront this crisis realistically? And still deeper in the background: Why are populations so willing to look the other way when survival of organized human life is literally at stake?

One answer was given by a participant in the remarkable Yellow Vest uprising in France.

The immediate cause of the uprising was President Emmanuel Macron's proposal in 2018 to raise fuel taxes with alleged environmental concerns, a move that would hit the poor and working people in rural areas particularly hard. But the background to the protests, more broadly, was the whole range of Macron "reforms" that benefited the rich while harming poor and working people. The participant, perhaps a committed environmentalist himself as many were, said that you are talking about "the end of the world" but we are concerned with "the end of the month." How are we to survive your "reforms"?

A fair question, which quickly became the slogan of the grassroots demonstrations sweeping Paris and much of the rest of the country. And a question that cannot be overlooked by the environmental movement.

Global warming has an abstract feel. Who understands the difference between 1.5°C and 2°C (2.7°F and 3.6°F respectively)—in contrast to having food to put on the table for your children tomorrow? True, there are more storms, heat waves, other disturbances today—but others doubtless can conjure up something like my own personal experience. I've lived through many hurricanes in Massachusetts, but none as fierce as those almost seventy years ago. So maybe Trump is right when he says the climate always changes—sometimes it's warmer, sometimes colder? It's easy to fall into that trap when your prime concern is putting food on the table tomorrow.

And why follow President Carter's gloomy prescription of turning down the thermostat and putting on a heavy sweater, and in general cutting back on our lifestyle while billions of Chinese and Indians—so we hear on Fox News—are pouring pollutants into the atmosphere with abandon?

Or consider the mineworker in West Virginia who was cheering at a Bernie Sanders rally until Sanders said that for any chance at decent survival, we must stop producing coal. No applause for that line. That would mean losing his job, and there's not much attraction in an alternative in the growing service industries or installing solar panels, which, other reasons aside, would mean losing his pension and health care, which were won in hard union struggles and are tied to employment. Lose your job, and you lose not only personal dignity but also the means of survival.

Here we come up against a fateful decision by US labor in the 1950s: to choose class collaboration, making deals with corporate management for wages and benefits while abandoning control of the workplace and broader social reforms. That decision by US labor leaders contrasted with the choice by the very same unions in Canada to fight for health care for the whole population, not just ourselves. The results are quite visible. Canada has a functioning health care system while the US is burdened by an international scandal, with costs about double those of comparable countries and relatively poor outcomes, thanks in no small part to the inefficiency, bureaucratization, and profit-seeking of the largely privatized US system.

By choosing class collaboration, US labor leaders left that mineworker and others like him at the mercy of the corporate owners, who were free to cancel the bargain. Which is what they have done, quite dramatically since the dawning of the neoliberal years. In 1978, UAW president Doug Fraser finally recognized that the business classes never abandon class war, even if labor leaders agree to do so. Fraser criticized the "leaders of the business community" for having "chosen to wage a one-sided class war in this country—a war against working people, the unemployed, the poor, the minorities, the very young and the very old, and even many in the middle class of our society," and for having "broken and discarded the fragile, unwritten compact previously existing during a period of growth and progress."[7]

Hardly a surprise, particularly in the US with its highly class-conscious business community and bitter history of violent suppression of labor, unusual in the developed world.

There followed years of neoliberal globalization designed in the interests of investors and the ownership class at the expense of American workers, alongside the neoliberal "reforms" guided by the same fundamental imperatives. The results should be well known. Wealth concentrated sharply, with the obvious consequences for functioning democracy, while real wages stagnated. Workers now have about the same purchasing power as they did forty years ago.[8] Unions came under bitter attack during the extreme anti-labor Reagan administration, a process carried forward under his successors. Demolition of the labor movement is a major achievement of neoliberal policies, following the Thatcherite doctrine that there is no society, only individuals, isolated creatures who face market discipline on their own, unorganized. That has been a core principle of neoliberalism going back to its Austrian origins in the 1920s. It's why the far-right "libertarian" guru, Ludwig von Mises, in the interest of preserving "sound economics" from interference, enthusiastically welcomed the crushing of the vibrant Austrian labor movement and social democracy by state violence in 1928, laying the groundwork for Austrian fascism; and in his major work, *Liberalism*, he praised fascism for saving European civilization.

To be sure, the atomization principle holds only for what Thorstein Veblen called "the underlying population." Those who matter, private wealth and corporate power, are highly organized in pursuit of their class goals, manipulating state power in their interests while the rest become "a sack of potatoes," to borrow Marx's phrase in his condemnation of the autocratic regimes of his day. The sack of potatoes, unorganized and increasingly consigned to precarious work and lives, are much more easily controlled.

Returning to the mineworker and many others like him, it is not hard to discern good reasons why they should resonate to the Yellow Vest slogan and resist the mass mobilization that is essential if we are to overcome the environmental crisis.

For organizers and activists, all of this provides important lessons. Revival of the labor movement is an essential task for many reasons. One is the environmental crisis. If the sack of potatoes becomes organized, active, and committed, it could become a leading force in the environmental movement. These are, after all, the people whose lives and future are at stake. It's not an idle dream. In the 1920s, the vigorous US labor movement had been crushed by state and business oppression, often through direct violence. The title of labor historian David Montgomery's classic *The Fall of the House of Labor* refers to that period. But a few years later, a lively and militant labor movement rose from the ashes and spearheaded the New Deal reforms that have greatly improved the lives of Americans through the great postwar growth period, before falling victim to the neoliberal assault. It's worth remembering that Bernie Sanders's revolution would not much have surprised Dwight Eisenhower, an outspoken supporter of New Deal measures.

It might be worth recalling the attitudes of the last conservative president, just to see how far we have come in the neoliberal age. Eisenhower declared:

> I have no use for those—regardless of their political party—who hold some foolish dream of spinning the clock back to days when unorganized labor was a huddled, almost helpless mass ... Only a handful of unreconstructed reactionaries harbor the ugly thought of breaking unions. Only a fool would try to deprive working men and women of the right to join the union of their choice ... Should any political party attempt to abolish social security, unemployment insurance, and eliminate labor laws and farm programs you would not hear of that party again in our political history. There is a tiny splinter group of course that believes you can do these things. Among them are ... a few ... Texas oil millionaires, and an occasional politician or businessman from other areas. Their number is negligible and they are stupid.[9]

They were in fact far from stupid. They were well-organized, committed, and waiting for the opportunity to show that "you can do these things," the basic thrust of the neoliberal age.

The revival of the labor movement in the '30s is an important precedent, but there are more recent ones. It's well to remember that one of the first and most prominent environmentalists was a union leader, Tony Mazzocchi, head of the Oil, Chemical and Atomic Workers International Union (OCAW). The members of his union were right on the front line, facing destruction of the environment every day at work, and were the direct victims of the corporate assault on individual lives. Under Mazzocchi's leadership, the OCAW was the driving force behind the establishment in 1970 of the Occupational Safety and Health Act (OSHA), protecting workers on the job, signed by the last liberal American president, Richard Nixon—"liberal" in the US sense, meaning mildly social democratic.

Mazzocchi was a harsh critic of capitalism as well as a committed environmentalist. He held that workers should "control the plant environment" while also taking the lead in combating industrial pollution.

By 1980, when it was clear that the Democrats had abandoned working people to their class enemy, Mazzocchi began to advocate for a union-based Labor Party. That initiative made considerable progress in the 1990s but couldn't survive the decline of the labor movement under severe business-government attack, reminiscent of the 1920s.[10]

The project could be revived, just as it has been in the past. Recent militant action in the growing service industries might be a harbinger of things to come, along with the impressive strikes by teachers in the red states, aimed not just at overcoming their miserable wages but more importantly at improving the woefully underfunded public school system—another target of the neoliberal assault on society. The path that Mazzocchi tried to forge—militant labor as a driving force of the environmental movement—is not an idle dream and should be actively pursued.

We have known about the effect of greenhouse gases since the mid-nineteenth century, and some scientists began warning us of the potential risks of a hotter planet decades ago, even while there are still some who deny that climate change is happening or that human activity is behind the phenomenon of global warming. But is it enough to point to human activity as the cause of global warming? Shouldn't we understand this crisis as resulting from the specific economic system that has been guiding economic life for the past five hundred years? And, if so, how exactly are capitalism and the climate crisis interconnected?

NC: There was no more enthusiastic cheerleader for the achievements of capitalism than Karl Marx, who did not, of course, fail to emphasize and explore its horrifying human and material consequences, in particular the "metabolic rift," a concept on which John Bellamy Foster has elaborated extensively: the inherent tendency of capitalism to degrade the environment that sustains life.[11]

In considering the impact of capitalism, and the options it may make available, it is worthwhile to bear in mind the actual nature of the systems to which this rather vague term is applied. In the spectrum of major state-capitalist societies (personally, I'd be inclined to include the USSR, but put that aside), the US is at the extreme end of capitalist orthodoxy. No other country so exalts what economist Joseph Stiglitz, twenty-five years ago, criticized as "the 'religion' that markets know best" (exalts in words at least; practice is a different matter). Consider, then, its economic system throughout its history and today—leaving aside the state role in emptying the national territory of the native scourge and stealing half of Mexico in a war of aggression, thus providing the US with historically unparalleled natural advantages.

The foundation of US economic development (and of Britain as well), was the most vicious system of slavery in human history, qualitatively different from anything that came before. It created "the empire of cotton" (Sven Beckert's apt term): the basis for manufacturing, finance, commerce. A rather severe intervention in the holy market. The story continues. The Hamiltonian system of high tariffs enabled industry to develop domestically, as the newly liberated colonies firmly rejected Adam Smith's recommendation to keep to sound economics, producing primary products and adopting superior British manufactures in accord with their comparative advantage. It was also helpful to take superior British technology in ways now bitterly condemned as "robbery" when others do it. With justice, economic historian Paul Bairoch describes the US as "the mother country and bastion of protectionism," well into the mid-twentieth century, when its economy had advanced so far beyond the rest that "free trade" seemed to be a good bargain—imitating what Britain had done a century earlier. From an extensive review, Bairoch concludes that "it is difficult to find another case where the facts so contradict a dominant theory [as the theory] concerning the negative impact of protectionism."[12]

Skipping a lot, the American system of mass production that amazed the world—quality control, interchangeable parts, Taylorism—was mostly developed in government armories and military installations. Moving on to the present, what is misleadingly called "the military-industrial complex," more accurately today's high-tech economy, is substantially the outcome of taxpayer-funded R&D extended through a creative, costly, and risky period, often for decades before it is handed over to private enterprise for adaptation to the market and profit. It's a system that might be called "public subsidy, private profit," taking many forms, including procurement and far more. That includes technology we use now, computers and the internet, but much more.

It is not quite that simple of course, and this barely skims the surface, but the relevant point for our discussion is that so-called capitalism can readily accommodate major initiatives of industrial policy, public subsidy, state initiative, and market interference, and has done so throughout its history. The implications for today's ecological crisis should be clear.

To return to the specific question: basic elements of capitalism, both ideological and institutional, lead directly to destruction of the basis of organized social life—if unconstrained. We see that dramatically every day.

Take the well-studied case of the huge energy conglomerate ExxonMobil. From the 1960s, its scientists were in the lead in revealing the extreme threat of global warming. In 1988, geophysicist James Hansen issued the first major public warning of the extent of the threat. ExxonMobil management reacted by initiating a program of denialism taking many forms: typically raising doubts, since outright denial is too easily refuted. That continues to the present. Recently ExxonMobil, along with the Koch brothers, filed a formal complaint with NASA objecting to its reporting that 97 percent of climate scientists agree on human-caused global warming. The 97 percent consensus is well established by very careful studies, but a crucial element of the denialist strategy has been to sow doubt about it, with no little success: only 20 percent of Americans realize that over 90 percent of climate scientists accept the overwhelming consensus.[13]

All of this is done with full knowledge that they are engaged in pure deceit, with severely malignant consequences.

Even more malignant than the denialism is practice. ExxonMobil is in the lead in expanding fossil fuel production. Unlike some other oil majors, it does not want to waste even small sums on sustainable energy: "In a March [2014] report on carbon risk to shareholders," the business press reports, "ExxonMobil (XOM) argued that its laserlike focus on fossil fuels is a sound strategy, regardless of climate change, because the world needs vastly more energy and the likelihood of significant carbon reductions is 'highly unlikely.'"[14]

In extenuation, it can be argued that ExxonMobil is only being more honest than its competitors in following capitalist logic. The same article reports the decision of Chevron to close its small and profitable sustainable energy projects because destroying the environment is more profitable. Others are not all that different. Royal Dutch Shell right now is celebrating the establishment of a huge plant to produce non-biodegradable plastic, in the certain knowledge that it will destroy the oceans.[15]

The same cynicism also prevails elsewhere in the ruling class. The CEO of JPMorgan Chase understands as much about the extreme threat of global warming as other educated people—and in private life may well be a contributor to the Sierra Club. But he has been pouring huge resources into developing fossil fuels, including the most dangerous of them, Canadian tar sands—also a favorite of the energy industries.

It's easy to expand the list. All are following impeccable capitalist logic, knowing exactly what the consequences are, but in a certain sense having no individual choice: if the CEO chooses otherwise, he will be replaced by someone who will do the same thing. The problem is institutional, not merely individual.

To this grim list we can add the regular euphoric articles in the finest journals on how fracking has propelled the US once again to the championship in production of the fossil fuels that will destroy us, achieving "energy independence"— whatever that is supposed to be—and providing the US with leverage to pursue its (by definition benign) international objectives without concern about energy markets, like seeking to impose maximal suffering on the people of Iran and Venezuela. Occasionally there are a few words about environmental consequences: fracking in Wyoming may harm water supplies for ranchers. But one will search in vain for a comment on what this means for the future of life.

Again, in extenuation, we must recognize that to refer to such side issues as human survival would violate the canon of "objectivity" and introduce "bias": the story assigned by the editors is fracking and its contribution to US dominance of fossil fuel production. So survival must be left to the rare opinion column. The effect of course is to instill more deeply the sense of "don't worry." If there's a problem, human ingenuity will figure out how to deal with it.

It might finally be worth noting that not only the management of major corporations but also the most extreme denialists are well aware of the impending disaster to which they are contributing. The capitulation to the Koch brothers a decade ago is one illustration. Or the president, who understands enough to appeal to the government of Ireland for permission to build a wall to protect his golf course from rising sea levels.[16] Some things matter.

We can add finally a prime candidate for the most astonishing document in human history, produced by the Trump administration in August 2018: a five-hundred-page environmental impact statement by the National Highway Traffic Safety Administration, which concluded that no new restrictions are needed for automotive emissions. The authors had a sound argument: their assessment concluded that by the end of the century temperatures will have risen 4°C above preindustrial levels, about twice the level that the scientific community regards as catastrophic. Automotive emissions are only one contributor to total catastrophe. Accordingly, since we are going off the cliff anyway in the near future, why not drive freely while the world burns, far outdoing Nero?[17]

If one can find a document of comparable malevolence in the historical record, I would be interested in knowing about it. Even the January 1942 Wannsee Conference of the Nazi leadership called only for the destruction of European Jewry, not of most human and animal life on Earth.

As usual, the study was released and circulated with virtually no comment.

The Trump administration argument, of course, assumes that the criminal insanity of the Republican Party leadership is shared universally, so that nothing will be done to avert catastrophe. But, putting aside the attitudes, for which there are no appropriate words in the language, what is relevant here is their clear recognition of what they are doing as they pull out all stops to increase the use of destructive fossil fuels and fill the overstuffed pockets of their prime constituency, wealth and private power.

In brief, capitalist logic, left unconstrained, is a recipe for destruction. However, a simple consideration of time scales reveals that the existential issues must be addressed within the framework of state-capitalist systems. These can accommodate radical market interferences and major state initiatives. To develop these options is one crucial task of social movements. And another, in parallel, is to undermine this logic at its roots and to prepare the ground for a sane society.

Opportunities abound. I already mentioned Tony Mazzocchi's initiatives. For them to have succeeded was within the bounds of realism, and remains so. And there are others. Let's try a thought experiment. Suppose that in 2008, when the Great Recession struck, a president had been in office who was not bound by strict capitalist logic; someone like Bernie Sanders, perhaps. Suppose further that the president had congressional support and was backed by activist popular movements. There were options. One would have been to honor the congressional legislation that provided taxpayer bailouts for the financial institutions who were responsible for the crash as well as relief for their victims who lost their homes. That possibility was dismissed: only the first commitment was considered worthy of fulfillment, a decision that infuriated Neil Barofsky, the special Treasury Department inspector general charged with overseeing the Troubled Asset Relief Program, or TARP. (Barofsky later wrote an angry book outlining the crime.)[18] Evidently, a different choice was possible.

But let's be more imaginative, while still not straying from the real world. When the crisis struck, Obama virtually nationalized the US auto industry. This major part of the US industrial system was in substantial measure in government hands. That step too opened up options. One, adopted reflexively, was to return the industry to the former owners and managers, perhaps under new names, who would then proceed as before to produce cars for profit. Another option would have been to turn the industry over to stakeholders, the workforce and community, effectively socializing a core part of the US industrial system. Perhaps, considering human life instead of mere profit, they might have decided to reorient production, realizing that efficient mass transportation yields a better life than spending hours a day fuming in traffic jams—and also alleviates the impending environmental threat in no small way.

Socializing a central part of the US industrial system in the real sense—placing it under worker and community control—would be a complex enterprise, with many facets, and would likely have large-scale effects beyond revitalizing the labor movement and inspiring other developments. Is that a utopian dream, beyond imagination? It doesn't seem so. Such opportunities arise constantly, even if on a lesser scale. In recent years worker-ownership and cooperative initiatives have been proliferating. The Next System Project, initiated by Gar Alperovitz, is coordinating and initiating many such efforts, establishing the basis for a future free and democratic society within the present one, Mikhail Bakunin's prescription.[19] And considerably larger goals can be realistically contemplated.

We should also not overlook the potential of popular activism and pressures. To mention a few examples from early 2020, in a report to clients that was leaked to the environmental activist organization Extinction Rebellion, JPMorgan Chase revealed deep concerns about climate warming. The bank, reported the *Guardian*, "warned clients that the climate crisis threatens the survival of humanity and that the planet is on an unsustainable trajectory [with] irreversible consequences" unless the trajectory changes. It also recognized its own investment strategies must change because of the "reputational risks" of fossil fuel investment.[20]

The phrase "reputational risks" refers to public pressures. Changing the investment strategies of "the world's largest financier of fossil fuels" would be no slight achievement.

To mention another case, the world's richest man, Jeff Bezos, announced in February that the new Bezos Earth Fund would provide $10 billion in grants to scientists and activists to fund their efforts to fight "the devastating impact of climate change on this planet we all share." His announcement came, according to the *Washington Post*, "one day before company employees—members of Amazon Employees for Climate Justice—planned to walk off the job in protest, saying the retailer and tech giant needs to do more to reduce its carbon footprint," and on the same day that PBS's *Frontline* was airing an investigation of the "Amazon Empire," examining the company's practices. Again, the result of public activism.[21]

There are many opportunities to have a meaningful impact on consciousness and practice.

What about the argument that it isn't really capitalism that should be blamed for the current climate crisis, but rather industrialization itself? After all, the environmental damage that the former so-called socialist world (the former USSR and Eastern Europe) caused in its rather short life is now well documented.

NC: I would emphasize the phrase "so-called." This is not the place to pursue the matter, but we should view with caution the claims of major propaganda systems. There have been two primary ones: the huge US system and its pathetic Eastern counterpart. They disagreed on many things, but not everything. They agreed that the radical perversion of socialism in Eastern Europe was "socialist"—the US in order to defame socialism; the Soviets, to try to benefit from some of socialism's moral aura. We're not compelled to follow suit.

The fact that capitalist logic, unconstrained, leads directly to destruction of the environment does not entail that it is the only possible source of this outcome. There is a great deal to say about the harsh and brutal process that turned Russia from a very poor peasant society, continuing to decline relative to the West as it had been doing for centuries, into a major industrial power, despite the terrible traumas of wars. But there is no escaping the fact that the environmental impact was devastating.

The Western mode of industrialization relied on slavery (creating "the empire of cotton" and the basis for much of the modern economy), coal (found in abundance in England, then elsewhere), and, in the twentieth century, oil. Was this a necessity? Were there other ways to develop an industrial society, perhaps of a very different kind, with radically different social and economic institutions and concern for the human and environmental impact of decisions and implementation? The question has not been extensively pursued, and answers don't seem obvious. Until they are investigated I don't think we can go so far as to cast the blame on "industrialization itself." There might well have been roads not taken, radically different ones.

Bob, how do you conceive of the relationship between capitalism and climate change?

RP: The rise of capitalism was certainly tightly bound up with burning fossil fuels to produce energy and power machines. Contaminating the atmosphere with CO_2 emissions was therefore also tied up with the emergence of industrial capitalism.

But this connection is not simply a matter of manufacturing capitalists needing energy sources in general to power machines as the industrial revolution emerged in the late 1700s and early 1800s. What really happened is that coal began being used intensively in Great Britain in the 1830s to power steam engines for producing cotton and then other manufacturing products. At that time, coal was in the process of supplanting waterpower as the primary energy source in manufacturing. As of around 1850, 60 percent of all global CO_2 emissions from fossil fuels were generated in Britain by burning coal.

However, as Andreas Malm has demonstrated in detail in his book *Fossil Capital* and elsewhere, the early nineteenth-century British manufacturers turned to coal and steam engines to supplant water wheels not because coal and steam provided a lower-cost, much less a cleaner, alternative to waterpower. Indeed, waterpower was less expensive at the time, and the technology for driving machines with waterpower was more advanced.

Rather, the overwhelming advantage of coal and steam power was that they were not bound to specific locations. Waterpower could only be provided adjacent to where powerful streams of water happen to have been situated.

With coal, the manufacturing operations could be located anyplace where coal could be delivered and burned. This made it much easier for businesses to get people to show up in the factories to work, since, as was well known, the working conditions were mostly abysmal. As Malm writes:

> When a manufacturer came across a powerful stream passing through a valley or around a river peninsula, chances were slim that they also hit upon a local population predisposed to factory labour; the opportunity to come and work at machines for long, regular hours, herded together under one roof and strictly supervised by a manager, appeared repugnant to most, and particularly in rural areas.[22]

By contrast, as Malm explains,

> Steam was a ticket to the town, where bountiful supplies of labour waited. The steam engine did not open up new stores of badly needed energy so much as it gave access to exploitable labor . . . an advantage large enough to outdo the continued abundance, cheapness and technological superiority of water.[23]

Providing capitalists with this newfound freedom to locate their manufacturing operations wherever they could lure an exploitable labor force into their factories became, in turn, a propulsive force for the expansion of capitalism beyond Britain's borders, into the rest of Europe, North America, and the colonies of the various European powers. Marx himself described this explosive growth of capitalism vividly in chapter 1 of *The Communist Manifesto*, writing: "The need of a constantly expanding market for its products chases the bourgeoisie over the entire surface of the globe. It must nestle everywhere, settle everywhere, establish connexions everywhere."[24] The manufacturing capitalists of Marx's era could not have been capable of nestling everywhere, settling everywhere, and establishing connections everywhere if they had remained location-bound, as they were with waterpower.

At the same time, we do know that, for both better and worse, capitalism can operate just fine in our present era without having to rely exclusively on coal, oil, and natural gas as energy sources. Workers are exploited in China, the US, Brazil, and Russia, among other places, by operating machines driven by hydroelectricity.

Yet it is also the case that the expansion of clean energy supply—primarily solar and wind power—is creating opportunities for smaller-scale enterprises, which could be organized through various combinations of public, private, and cooperative ownership structures—that is, a variety of capitalist, non-capitalist, and mixed ownership structures. The performance of these non-corporate business enterprises has generally been quite favorable relative to traditional corporate firms. One area where this has been clearly demonstrated is community-based wind farms in Western Europe, especially in Germany, Denmark, Sweden, and the United Kingdom. Variations on this wind farm model are also emerging in the US Midwestern farm belt. Private farmers, large and small, are siting wind turbines on their crop-growing and cattle-grazing farmland. This second use for their farmlands provides an additional, and often significant, income source for the farmers.

In short, global capitalism did indeed emerge as Malm has vividly described, on the foundation of a fossil-fuel-based energy system. It is also possible that the necessary clean energy transition could provide one critical cornerstone for advancing more democratic, egalitarian societies. But we should be 100 percent clear that this outcome is by no means guaranteed. No technology, either clean energy or anything else, can, by itself, deliver meaningful social transformations. Egalitarian social transformations only happen when people effectively struggle to build political movements. When such political movements do emerge, technologies such as clean energy can then certainly play a critically important supportive role.

Capitalism is all about profits, and fossil fuels constitute the energy source that feeds the beast. Aren't capitalist profits at stake if efforts were pursued to shift energy resources away from fossil fuels?

RP: It is certainly the case that the profits of private *fossil fuel capitalists* are at stake. Indeed, fossil fuel companies will need to be either put out of business altogether, or at least dramatically diminished, within the next thirty years. According to the best estimates currently available, the reserves of "unburnable" oil, coal, and natural gas in the ground that these private companies now own amount to about $3 trillion. These reserves can never be burned and thus converted into capitalist profits if the planet is going to have a decent chance of stabilizing the climate.

Of course, the fossil fuel companies will fight by all means available to them for the right to profit lavishly from selling this oil, coal, and natural gas still in the ground. But it is also important to understand that it will not present a major problem for the rest of the global economy if the fossil fuel companies are indeed prevented from selling their $3 trillion in unburnable assets. Let me illustrate this point with a simple numerical example. Yes, of course, $3 trillion is a huge sum of money. But, as of 2019, it equals less than 1 percent of the $317 trillion in total worldwide private financial assets—the total value of all equity and debt assets outstanding.[25] Still more, the anticipated $3 trillion decline in the value of private fossil fuel assets will not happen in one fell swoop, but rather will occur incrementally over a thirty-year period. On average, this amounts to asset losses of $100 billion per year, or 0.03 percent of the current value of the global financial market. By contrast, as a result of the US housing bubble and subsequent financial collapse in 2007–9, US homeowners lost $16 trillion in asset values in 2008 alone—about 160 times the annual losses fossil fuel companies would face.

The fact that the decline in fossil fuel asset values will occur incrementally over two to three decades also means that the shareholders who own the fossil fuel companies will have ample opportunity to sell their stocks and move their money into other stocks. As one important example, in 2014 Warren Buffett, the best-known investor and third-richest person in the world, announced that his holding company, Berkshire Hathaway, was doubling its holdings in solar and wind energy companies to $15 billion. This is even while Berkshire continues to hold large positions in conventional utility companies.[26]

The fossil fuel companies could themselves follow the Buffett example by diversifying into clean energy. In fact, they are already doing so, if you believe their advertising campaigns. But the reality is that their forays into clean energy still represent a tiny fraction of their overall operations. Over decades, these companies have built up a capacity to earn super profits from producing and selling fossil fuel energy. They are not likely to achieve comparable levels of profitability with clean energy, because solar and wind technologies can generate power at a much smaller scale than fossil fuel technology. We know, for example, that average homeowners most anywhere in the world can right now generate 100 percent of their entire electricity needs and save themselves money by putting solar panels on their roofs. Over time, the fossil fuel companies will have no way to compete against that.

This brings up a more general point. The profits of the fossil fuel companies are most certainly at stake through a clean energy transformation, as are those of ancillary industries, such as oil drilling and pipeline construction companies, the railroad companies that transport coal, and all the utilities that now burn fossil fuels to generate electricity. But there is no reason to expect that other capitalist enterprises should see their profits threatened because they have to rely on solar or wind power for their energy supply instead of oil, coal, and natural gas. Electricity generated by onshore wind or solar photovoltaic panels is already at approximate cost parity with electricity generated by coal or natural gas. The costs of clean energy should also continue falling as the technologies come into more widespread use. Especially after the past forty years of massively rising inequality under neoliberalism, there is every justification for capitalists' profits to be pushed way down. But a clean energy transformation will not deliver this outcome on its own.

While there is no reason to think that capitalism cannot make a transition to clean energy resources, the fact of the matter is that short-termism guides the actions of most investors in the neoliberal age, so isn't it a bit naive to rely on capitalists themselves to get us out of the climate crisis?

RP: To be fair, nobody really believes that capitalists on their own will get us out of the climate crisis. Even the long list of prominent orthodox economists that I referred to above who signed the January 2019 statement supporting a carbon tax are clear that government intervention is necessary to force capitalists to integrate the costs of ecological destruction into their calculations. That is exactly the idea behind their support for a carbon tax.

The real question then is: *To what extent* is public intervention in the normal operations of capitalist markets needed to mount a successful global climate stabilization project? In my view, as I discussed above, this will require much more forceful forms of government intervention than the carbon tax, certainly considering the carbon tax is a standalone policy. In fact, we also need public investments in the critical economic sectors, public subsidies for private green investments as well as strong regulations. This combination of policies will be capable of moving us off of fossil fuels much more quickly than what is likely through relying on a carbon tax–type intervention by itself. If we look back at the mobilization project to fight World War II, the federal government did not just intervene through adjusting the tax system. The circumstances then called for much stronger measures, as they do now. Thus, as Josh Mason and Andrew Bossie show in a recent paper, during World War II, the Roosevelt government assumed a major role in the areas of public investment and ownership of the most critical industries. This included 97 percent of the synthetic rubber industry, 89 percent of the aviation industry, 87 percent of shipbuilding, and 14 percent of even such an established industry as iron and steel.[27] The Roosevelt government took over these industries because it was clear that, on their own, private capitalists were not about to assume the risks of raising production levels at either the speed or scale that the crisis warranted.

Our situation today is comparable. This is why, as I mentioned above, I am convinced that a viable Green New Deal for today needs to include substantial levels of public investment, public ownership, and hard-cap regulations. As one example, if some electric utilities are going to remain privately owned, then they must commit to reducing their level of CO_2 emissions every year by set amounts that will ensure they reach the zero emissions target by 2050. The CEOs of the companies should then face jail time if they fail to meet these requirements. I flesh out these ideas in more detail in section 3, on the Green New Deal.

Noam, what are your thoughts on this matter? Can we realistically expect the current economic system, with profit-making as its driving force, to rescue humanity and the planet as a whole from the potentially catastrophic effects that lie ahead if we fail to contain the menace of global warming?

NC: If profit-making remains the driving force, then we are doomed. It would be the sheerest accident, too remote to consider, for pursuit of profit to somehow magically lead to the termination of such highly profitable activities as producing fossil fuels, or even far lesser forms of destruction. When we look closely, we commonly find that market signals are either wholly inadequate or are leading in entirely the wrong direction. To take just one current case, developing technology to remove carbon from the atmosphere is of prime importance, but for venture capitalists in Silicon Valley, investing in long-term projects with no likely major profits is far less attractive than adding new bells and whistles to iPhones.

The worship of markets is by now part of Gramscian hegemonic common sense, instilled by massive propaganda, particularly during the neoliberal years. The "religion," to borrow Stiglitz's term, is based on a particular view of human nature that is hardly compelling, to put it mildly. Do we really believe that humans would prefer to vegetate unless driven to action by profit? Or could it be, as a long tradition holds and ample experience reveals, that meaningful and creative work under one's own control is one of the joys of life?

In fact, it is highly misleading to say that pursuit of profit has been the main driving force in the past, even in the domain of industrial production. Consider again what we are now using, computers and the internet, developed for decades primarily within the state-university system before the results of this creative work were handed over to private enterprise for marketing and profit. For the most part, the driving motive of those who carried out the essential work was not profit but rather curiosity and the excitement of solving hard, challenging, and important problems. That's commonly true of other research and inquiry on which the health of our society and culture has relied. True, what was created was integrated into the profit-driven economic system, but that is not a law of nature. Society could be constituted differently. Worker-owned and -managed enterprises, for example, can be expected to have different priorities than profit for bankers in New York—decent working conditions and ample room for individual initiative and leisure, for example. And if those enterprises are linked together, and to truly democratic communities, something quite different might emerge: perhaps shared values of mutual aid and concern for a meaningful and fulfilling life rather than amassing commodities for oneself and enriching those who have capital to invest.

Can we realistically expect this? We don't know. What will be "realistic" depends in part on our choice of action.

Having focused on the interconnectedness between capitalism and the climate crisis, we should not also forget that, in many cases, fossil fuel industries are publicly owned enterprises, which makes one wonder about the role of public entities under capitalism. Bob, how should we think of public entities and their contribution to the climate crisis?

RP: In fact, throughout the world, the energy sector has long operated under a variety of ownership structures, including public/municipal ownership and various forms of private cooperative ownership, in addition to private corporate entities. Indeed, in the oil and natural gas industry per se, publicly owned national companies control approximately 90 percent of the world's reserves and 75 percent of production. They also control many of the oil and gas infrastructure systems. These national corporations include Saudi Aramco, Gazprom in Russia, China National Petroleum Corporation, the National Iranian Oil Company, Petroleos de Venezuela, Petrobras in Brazil, and Petronas in Malaysia. None of these publicly owned companies operates with the same profit imperatives as big private energy corporations such as ExxonMobil, British Petroleum, and Royal Dutch Shell. But this does not mean that they are prepared to commit to fighting climate change simply because we face a global environmental emergency. Just as with the private companies, producing and selling fossil fuel energy generates huge revenue flows for these companies. National development projects, lucrative careers, and political power all depend on continuing the flow of large fossil fuel revenues. We should therefore not expect that public ownership of energy companies will, by itself, provide a more favorable framework for advancing effective clean energy industrial policies.

3

A Global Green New Deal

Progressive economists and environmentalists alike have been proposing over the years the introduction of zero-emission energy resources to stave off the effects of climate change. The adoption of clean, renewable sources of energy is often a central plank of what advocates call a "Green New Deal," a bold vision of environmental economics, one might say, inspired by FDR's New Deal and guided largely by the logic of Keynesian economics regarding growth. Still, there are different versions of a Green New Deal that have been proposed by different people, so the critical question is what constitutes a realistic and sustainable project for achieving zero-emission energy resources by 2050 that can overcome existing political, economic, and even cultural resistance to a "green economy." Bob, you have already done extensive work over the past decade in advocating on behalf of a Green New Deal. What then constitutes for you a politically realistic and economically feasible Green New Deal project? What does it entail, and how will it work?

RP: The IPCC estimates that, to achieve the 1.5 degrees maximum global mean temperature increase target as of 2100, global net CO_2 emissions will have to fall by about 45 percent as of 2030 and reach net zero emissions by 2050. As such, by my definition, the core of the global Green New Deal is to advance a global project to hit these IPCC targets, and to accomplish this in a way that also expands decent job opportunities and raises mass living standards for working people and the poor throughout the world. It is that simple.

In fact, purely as an analytic proposition and policy challenge—independent of the myriad of political and economic forces arrayed around these matters, which we take up later—it is entirely realistic to allow that global CO_2 emissions can be driven to net zero by 2050. By my higher-end estimate, it will require an average level of investment spending throughout the global economy of about 2.5 percent of global GDP per year, focused in two areas: (1) dramatically improving energy efficiency standards in the stock of buildings, automobiles and public transportation systems, and industrial production processes; and (2) equally dramatically expanding the supply of clean renewable energy sources—primarily solar and wind power—available to all sectors and in all regions of the globe, and at competitive prices relative to fossil fuels and nuclear power. These investments will also need to be complemented in other priority areas, the most important of which, as I discussed earlier, are stopping deforestation and supporting afforestation.

Focusing on the clean energy transformation, the level of necessary investments would amount to about $2.6 trillion in the first year of the program. To be realistic, I assume that the project won't begin in earnest until 2024. Spending would then average about $4.5 trillion per year between 2024 and 2050. Total clean energy investment spending for the full twenty-seven-year investment cycle would amount to about $120 trillion.

These figures are for overall investment spending, including from both the public and private sectors. Establishing the right mix between public and private investment will be a major consideration within the framework of industrial and financing policies. As we have discussed above, it is certainly not realistic to expect that this can all be accomplished through private capitalist investments. But it is equally unrealistic to expect that public enterprises, on their own, can mount a project at this scale, and with the speed that is required. Still, advancing the Green New Deal will itself be a major force driving the transformation of capitalism away from its current interregnum between neoliberalism and neofascism. As I discussed above, the Green New Deal will create major new opportunities for alternative ownership forms, including various combinations of smaller-scale public, private, and cooperative ownership. A major reason why these enterprises have succeeded in Western Europe is that they operate with lower profit requirements than those of big private corporations. Given all this, we will still also need to bring big private capitalist firms into the mix, though they will have to be heavily regulated.

As for the details of the program, I think it is reasonable to assume that global clean energy investments should be divided roughly equally—that is, 50 percent public and 50 percent private investment respectively on a global basis. For the first year of full-scale investment activity in 2024, this would break down to $1.3 trillion in both public and private investments. A major part of the policy challenge will be to determine how to leverage the public money most effectively to create strong incentives for private investors, large and small, while also maintaining tight regulations over their activities.

It is important to emphasize that this clean energy investment project, the centerpiece of the Green New Deal, will pay for itself in full over time. More specifically, it will deliver lower energy costs for energy consumers in all regions of the world. This will result from the new energy efficiency standards, which will ensure that consumers spend less for a given energy-intensive activity. Compare, in the case of driving, being able to travel 100 miles on a gallon of gasoline with a high-efficiency hybrid plug-in vehicle with getting only 25 miles per gallon with the average US car.[1] Moreover, the costs of supplying energy through solar and wind power, as well as geothermal and hydro, are now, on average, roughly equal to or lower than those for fossil fuels and nuclear energy. As such, the initial up-front investment outlays can be repaid over time through the forthcoming cost savings.

For 2018, the level of global clean energy investment, including both energy efficiency and clean renewable investments, was at about $570 billion, equal to about 0.7 percent of the current global GDP level of about $86 trillion. Thus, to meet the IPCC targets, the increase in clean energy investments will need to be in the range of 1.8 percent of global GDP—that is, about $1.5 trillion at the current global GDP level, then rising in step with global GDP growth thereafter until 2050.

The consumption of oil, coal, and natural gas will also need to fall to zero over this same thirty-year period. The rate of decline can begin at a relatively modest 3.5 percent in the initial years of the transition program, but then will need to increase every year in percentage terms, as the base level of fossil fuel supply contracts to zero by 2050. Noam and I have both discussed this above, but it bears repeating: of course, both the privately owned fossil fuel companies, such as ExxonMobil and Chevron, and equally, the publicly owned companies such as Saudi Aramco and Gazprom in Russia, have massive self-interests at stake in preventing reductions in fossil fuel consumption as well as enormous political power. These powerful vested interests will simply have to be defeated. How exactly we accomplish this is, of course, the most challenging question at hand. But it absolutely must be done. We return to this critical issue later.

There are also significant technical challenges that we will need to overcome to get to a zero emissions global economy by 2050. These concern the questions of land use requirements for installing an adequate supply of solar panels and wind turbines to meet global energy demand, as well as the combined issues of intermittency, transmission, and storage. "Intermittency" refers to the fact that the sun does not shine and the wind does not blow twenty-four hours a day. Moreover, on average, different geographical areas receive significantly different levels of sunshine and wind. As such, the solar and wind power that are generated in the sunnier and windier areas of the globe will need to be stored and transmitted at reasonable costs to the less sunny and windy areas.

The issues around transmission and storage of wind and solar power will not become pressing for many years into the clean energy transition, probably for at least a decade. This is because fossil fuels and nuclear energy will continue to provide a baseload of non-intermittent energy supply as these energy sectors proceed toward winding down while the clean energy industry rapidly expands. After all, fossil fuels and nuclear energy now provide roughly 85 percent of all global energy supplies. These supplies are not going to be eliminated overnight. Meanwhile, my understanding is that fully viable solutions to these technical challenges with transmission and storage of solar and wind power—including with respect to affordability—should not be more than a decade away, certainly as long as the market for clean energy grows rapidly at the required rate.[2]

A related issue is whether there will be sufficient supplies of the full set of raw materials that will be needed to rapidly expand the renewable energy sector. The short answer is "yes." Some short-term bottlenecks are likely to emerge for some of the required materials, in particular tellurium, which is used to produce solar cells. But none of the likely shortages, including with tellurium, should be insurmountable. One solution will be to greatly expand the industry for recycling the needed metals and minerals. At present, average recycling rates for these resources are below 1 percent of total supply. Increasing recycling rates to only 5 percent will go far to overcome any problems regarding supply shortages.[3]

In addition to recycling, opportunities will also emerge to economize on the level of minerals and metals necessary to produce solar panels, wind turbines, and batteries, as production technologies improve along with the rapid expansion of the industry. Substitute materials can also be developed for those materials that remain in short supply. What happened in recent years with neodymium, a metal used for producing wind turbines and electric vehicles, provides a valuable case in point. When the world price of neodymium peaked in 2010, producers found ways to economize on its use or eliminate it altogether as a necessary material. Demand for neodymium rapidly fell by between 20 and 50 percent as other materials were found to be adequate substitutes.[4]

The issue of land use requirements is frequently cited to demonstrate that building a 100 percent renewable energy global economy is wildly unrealistic. The late Cambridge University engineer David MacKay provided the most detailed arguments on the heavy land use requirements associated with renewable energy in his 2009 book *Sustainable Energy without the Hot Air*. MacKay's arguments have been repeated frequently ever since. Thus, in a 2018 *New Left Review* article, Troy Vettese writes, "A fully renewable system will probably occupy one hundred times more land than a fossil-fuel-powered one. In the case of the US, between 25 and 50 percent of its territory, and in cloudy, densely populated countries such as the UK and Germany, all of the national territory might have to be covered in wind turbines, solar panels and biofuel corps to maintain current levels of energy production."[5]

Vettese provided virtually no evidence to support his claims, and in fact, they cannot be supported. This becomes clear through reviewing the work of the Harvard physicist Mara Prentiss. Prentiss's 2015 book *Energy Revolution: The*

Physics and the Promise of Efficient Technology, as well as her more recent follow-up discussions, demonstrates how the US economy could run entirely on clean renewable energy sources by 2050 or earlier. Her arguments can be readily generalized to the global economy.

Prentiss shows that well below 1 percent of the total US land area would be needed through solar and wind power to meet 100 percent of the country's energy needs. Most of this land use requirement could be met, for example, by placing solar panels on rooftops and parking lots, then operating wind turbines on about 7 percent of current agricultural land. Moreover, the wind turbines can be sited on existing operating farmland with only minor losses of agricultural productivity. Farmers welcome this dual use of their land, since it provides them with a major additional income source. At present, the states of Iowa, Kansas, Oklahoma, and South Dakota all generate more than 30 percent of their electricity supply through wind turbines. The remaining supplemental energy needs could then be supplied by geothermal energy, hydropower, and low-emissions bioenergy. This scenario includes no further contributions from solar farms in desert areas, solar panels mounted on highways, or offshore wind projects, among other supplemental renewable energy sources, though all of these options are viable possibilities, if handled responsibly.

It is true that conditions in the United States are more favorable than those in some other countries. Germany and the UK, for example, have population densities seven to eight times greater than the US and also receive less sunlight over the course of a year. As such, these countries, operating at high efficiency levels, would need to use about 3 percent of their total land area to generate 100 percent of their energy demand through domestically produced solar energy. Using cost-effective storage and transmission technologies, the UK and Germany could also import energy generated by solar and wind power in other countries, just as, in the United States, wind power generated in Iowa could be transmitted to New York City. Any such import requirements are likely to be modest. Both the UK and Germany are already net energy importers in any case.

In applying Prentiss's calculations for the US to the global economy, the most critical point is that in terms of both population density and the availability of sunlight and wind to harvest, average conditions are much closer to those in the US than to Germany and the UK. In a 2019 article, Prentiss also explains how, through a range of approaches, including battery storage and straightforward improvements in energy transmission systems, "science and technology are not preventing us from achieving a 100 percent US renewable energy economy."[6]

Noam, you are also an avid supporter of the Green New Deal idea. Is this a project to save the planet or to save capitalism, as some of the critics to the left of the political spectrum might argue?

NC: It depends on how it is developed. And notice that these are not alternatives. Some form of a Green New Deal is essential to "save the planet." What form? The best answer I know is the one that Bob has developed in considerable detail and describes briefly above. If successful, a Green New

Deal might also "save capitalism" in the sense that it will abort the suicidal tendencies of "really existing capitalism" and lead to some viable form of social organization that might fall within the very loose spectrum of what can be called "capitalism." Personally, I hope it would go well beyond, and think that such aspirations are not unrealistic, but that's a different matter.

One of the climate change topics that are increasingly being discussed more widely is the potential role that the new disruptive technologies can play in cleaning up the air from accumulated carbon dioxide. These include radical technological fixes such as geo-engineering. What is your view on such carbon-negative technologies?

RP: Negative emissions technologies include a broad category of measures whose purpose is either to remove existing CO_2 or to inject cooling forces into the atmosphere to counteract the warming effects of CO_2 and other greenhouse gases. One broad category of removal technologies is carbon capture and sequestration. A category of cooling technologies is stratospheric aerosol injections.

Carbon capture technologies aim to remove emitted carbon from the atmosphere and transport it, usually through pipelines, to subsurface geological formations, where it would be stored permanently. One straightforward and natural variant of carbon capture technology is afforestation. This involves increasing forest cover or density in previously non-forested or deforested areas with "reforestation"—the more commonly used term—as one component.

The general class of carbon capture technologies have not been proven at a commercial scale, despite decades of efforts to accomplish this. A major problem with most of these technologies is the prospect for carbon leakages that would result under flawed transportation and storage systems. These dangers will only increase to the extent that carbon capture becomes commercialized and operates under an incentive structure in which maintaining safety standards will reduce profits.

By contrast, afforestation can make a valuable contribution within a broader set of emissions reduction measures. This is because forested areas naturally absorb CO_2 to a significant extent. However, the big question with afforestation is, realistically, how large its impact can be in terms of absorbing the CO_2 already accumulated in the atmosphere or offsetting newly generated emissions produced by ongoing fossil fuel consumption. In a careful recent analysis by Mark Lawrence and colleagues at the Institute for Advanced Sustainability Studies in Potsdam, Germany, the authors conclude that afforestation could realistically reduce CO_2 levels by between 0.5 and 3.5 billion tons per year through 2050.[7] But, as noted above, current global CO_2 emissions levels are at 33 billion tons. If this estimate by Lawrence and coauthors is even approximately correct, it follows that afforestation can certainly serve as a complementary intervention within a broader clean energy transition program, but cannot bear the major burden of getting to zero emissions by 2050.

The idea of stratospheric aerosol injections builds from the results that followed from the volcanic eruption of Mount Pinatubo in the Philippines in 1991. The eruption led to a massive injection of ash and gas, which produced

sulfate particles, or aerosols, which then rose into the stratosphere. The impact was to cool the earth's average temperature by about 0.6°C for fifteen months.[8] The technologies
being researched now aim to artificially replicate the impact
of the Mount Pinatubo eruption through deliberately inject-
ing sulfate particles into the stratosphere. Some researchers
contend that to do so would be a cost-effective method of
counteracting the warming effects of greenhouse gases.

The study by Lawrence and his coauthors reviews all the
major negative emissions technologies: carbon capture, aerosol injections, and afforestation. Their overall conclusion
from this review is that none of these technologies is presently at a point at which it can make a significant difference
in reversing global warming. They write:

> Proposed climate geoengineering techniques cannot be
> relied on to be able to make significant contributions . . .
> Even if climate geoengineering techniques were actively
> pursued, and eventually worked as envisioned on global
> scales, they would very unlikely be implementable prior
> to the second half of the century . . . This would very
> likely be too late to sufficiently counteract the warming
> due to increasing levels of CO_2 and other climate forcers
> to stay within the 1.5°C temperature limit—and proba-
> bly even the 2°C limit—especially if mitigation efforts
> after 2030 do not substantially exceed the planned efforts
> of the next decade.[9]

This conclusion reached by Lawrence and his coauthors is
exactly in line with that of Raymond Pierrehumbert, cited
earlier by Noam. As we saw above, Pierrehumbert, a lead

coauthor of the Third Assessment Report of the IPCC, is emphatic in his 2019 paper, "There Is No Plan B for Dealing with the Climate Crisis," that geo-engineering does not offer a viable solution to the climate crisis.

Noam, what's your own view on this matter? Should we be also exploring the possibility of applying technological solutions to combat the climate crisis facing the planet?

NC: I don't claim the technical competence to make a truly informed judgment, but even lacking that, there is ample reason to regard radical forms of geo-engineering as a last resort to be undertaken if humans persist in denying what is before their eyes. The Pierrehumbert article I mentioned earlier gives a careful and judicious analysis of the technical options, and their severe limitations. He also makes it clear that there is every reason to pursue technological solutions insofar as they are feasible and potentially effective and we have reasonable confidence that they will not be harmful— not a trivial matter, because there are not only inevitable uncertainties but also constant tradeoffs. Thus, there's wide agreement on the need to move toward electrification— which requires copper, a wasting resource and one that under current technology at least can be mined only in ways that are ecologically quite harmful. Such conundrums are hard to avoid, but that is not a reason not to explore aggressively the kinds of technology that seem best suited to progress toward a sustainable and healthy ecosystem. Geo-engineering in fact covers a broad range of choices. In its general sense, humans have been engaged in it for a long time—artificial nitrogen fixation for fertilizers, for example. These practices are not easy to avoid, due to the greater demands for agricultural productivity tied to rising population levels and more available land being consumed by urbanization. But, as Bob discussed earlier, they also have well-known unacceptable effects unless carefully integrated into progressive forms of land management, which should be supplemented by technologies to remove carbon from the atmosphere, also a form of geo-engineering.

There is far more to be done. Industrial meat production, even aside from ethical considerations, should not be tolerated because of its substantial contribution to global warming. We have to find ways to shift to plant-based diets derived from sustainable agricultural practices, which aren't trivial tasks.

More generally, the entire socioeconomic system based on production for profit, with its inherent imperative for growth whatever the consequences, cannot be sustained. And there are fundamental issues of value that cannot be overlooked. What is a decent life? Should the master-servant relation be tolerated? Should one's goals be the maximization of commodities—the imperative drilled into consciousness by huge industries devoted to fabricating wants, features of modern society explored long ago by Thorstein Veblen? Surely there are higher and more fulfilling aspirations.

Is there room for nuclear energy in a future zero emissions economy?

RP: As of 2018, nuclear power provided about 5 percent of total global supply. Nearly 90 percent of global nuclear power supply is generated in North America, Europe, China, and India. In terms of the world reaching a net zero CO_2 emissions target by 2050, nuclear power provides the important benefit that it does not generate CO_2 emissions or air pollution of any kind while operating.

This is why there are strong advocates for massively expanding nuclear energy supply to help build a zero emissions global economy. These advocates include James Hansen, formerly of NASA. For decades, Hansen has been the world's best-known climate scientist fighting for decisive action on climate change. In 2015, Hansen, along with fellow prominent climate scientists Kerry Emanuel, Ken Caldeira, and Tom Wigley, wrote:

> The climate system cares about greenhouse gas emissions—not about whether energy comes from renewable power or abundant nuclear power. Some have argued that it is feasible to meet all of our energy needs with renewables. The 100 percent renewable scenarios downplay or ignore the intermittency issue by making unrealistic technical assumptions . . . Large amounts of nuclear power make it much easier for solar and wind to close the energy gap.[10]

Hansen's position was strongly endorsed in the 2019 edition of the *World Energy Outlook*, published by the International Energy Agency, the most widely recognized source on global energy issues. The 2019 *Outlook* concludes that "alongside renewable energy and [carbon capture] technologies, nuclear power will be needed for clean energy transitions around the world."[11]

However, to make this case for nuclear energy, these advocates have to downplay the range of fundamental problems that would be unavoidably attached to any large-scale global build-out of nuclear reactors. These problems, of course, start with issues of environmental impacts and public safety, including the following:

Radioactive wastes. These wastes include uranium mill tailings, spent reactor fuel, and other wastes, which according to the US Energy Information Administration "can remain radioactive and dangerous to human health for thousands of years."[12]

Storage of spent reactor fuel and power plant decommissioning. Spent reactor fuel assemblies are highly radioactive and must be stored in specially designed pools or specially designed storage containers. When a nuclear power plant stops operating, the decommissioning process involves safely removing the plant from service and reducing radioactivity to a level that permits other uses of the property.

Political security. Nuclear energy can obviously be used to produce deadly weapons as well as electricity. Thus, the proliferation of nuclear capacity creates dangers of this capacity being acquired by organizations, governments or otherwise, that would use that energy as instruments of war or terror.

Nuclear reactor meltdowns. An uncontrolled nuclear reaction at a nuclear plant can result in widespread contamination of air and water with radioactivity for hundreds of miles around a reactor.

For decades, the prevalent view throughout the world was that these risks associated with nuclear power were relatively small and manageable, when balanced against its benefits. However, this view was upended in the aftermath of the March 2011 nuclear meltdown at the Fukushima Daiichi power plant in Japan, which resulted from the massive 9.0 Tōhoku earthquake and tsunami. While the full effects of the Fukushima meltdown remain uncertain, the most recent estimate of the total costs of decommissioning the power plant and providing compensation to victims is $250 billion.[13]

Clearly, the safety regulations at Fukushima failed miserably to do their job. And remember that this happened in Japan, a high-income economy, and the country that has suffered from the effects of nuclear power more than anyplace else. If the Japanese nuclear safety regulations proved to have been a failure, why should we expect that much stronger and more effective regulations will be enforced elsewhere in the world under a massive build-out of nuclear reactors? Presumably this massive build-out would include countries with much tighter public safety budgets than Japan's.

This then raises the more general issue of costs. In fact, according to the Trump administration's own Energy Department, the costs of generating electricity from nuclear energy are now about 30 percent higher than those from solar or onshore wind.[14] Moreover, the costs of renewables, especially solar, have been falling sharply over the past decade, with further large cost reductions likely. By contrast, nuclear is on a "negative learning curve"—meaning the costs of nuclear energy have been rising over time, in large part, though not entirely, due to the increased understanding that to truly minimize the risks of another Fukushima-like disaster entails billions of dollars in additional costs to bring a single new reactor online. This is why the huge multinational firm Westinghouse, which for decades had been the global leader in building nuclear plants, was forced to file for bankruptcy in 2017.

Even the International Energy Agency, in advocating for nuclear in its 2019 *World Energy Outlook*, could mount only a feeble case in its behalf. It estimates that, if the advanced economies were to forego nuclear energy altogether as a component of their clean energy transitions, concentrating instead on renewables, the result would be "5 percent higher electricity bills for consumers in advanced economies."[15] As a worst-case scenario, something like a 5 percent increase in electricity prices is clearly a trivial amount to pay to avoid all the fully understood costs and dangers that are unavoidably tied to nuclear energy.

In advancing the global clean energy transition over the next thirty years, there can be a case for allowing the existing nuclear power plants that are well functioning to continue operating through the course of their normal service lives. But continuing to operate these existing plants for the next one to two decades cannot be confused with a massive expansion of new reactors, when we know that investments in energy efficiency and renewables can deliver a zero emissions global economy within thirty years.

NC: I frankly don't know. The dangers are apparent and familiar. With current technology, having nuclear power is not very far from the capacity to develop nuclear weapons, a true nightmare. And there are technical problems that have not been resolved, like disposal of nuclear wastes. I hope there is a way out of the present crisis without resort to nuclear energy, but I don't feel that the option should be simply foreclosed.

There's a growing literature linking climate change to economic inequality. What exactly is the link between climate change and inequality, and how would a Green New Deal help to reduce economic inequalities on a global scale?

RP: There are several ways in which climate change and inequality interact. We should begin with the question: Who is responsible for causing climate change, or, more specifically, who is responsible for putting the greenhouse gases into the atmosphere that are causing climate change? The short answer to this question is that, if we focus on CO_2 emissions, and trace back the burning of fossil fuels over the full industrial era—that is, roughly from 1800 to the present—then virtually the entire blame for causing climate change falls on the US and Western Europe. These are the regions of the world that, through at least 1980, were responsible for nearly 70 percent of all cumulative emissions. Considered on a per capita basis, the discrepancy in the contributions through 1980 is even more extreme. For example, as of 1980, the average annual emissions in the United States was about 21 tons per capita, fourteen times greater than the 1.5 tons per capita figure for China that year and forty-two times greater than India's figure of 0.5 tons per capita.[16]

But even these comparisons by country do not give us a full picture of the relationship between emissions levels and inequality. This is because, of course, the average levels of fossil fuel energy consumption, and thereby emissions, within any given country are also highly unequal according to people's income and overall consumption levels. Considering the global population as a whole by income levels, as of 2015, the richest 10 percent of the global population was responsible for nearly half of all emissions tied to personal consumption, while the poorest 50 percent of the population was responsible for only 10 percent of total consumption-based emissions.[17]

It is true that, with China having experienced a historically unprecedented pace of economic growth since the early 1980s, it is now the largest emitter of CO_2 emissions, at 9.8 billion tons in 2017 (27 percent of global emissions), compared with the United States at 5.3 billion tons (15 percent of global emissions). However, even here, on a per capita basis for 2017, China's emissions, at 7.0 tons per capita, are still less than half the US figure of 16.2 tons per capita.

A second critical way in which we need to think about inequality and climate change is in terms of impact: Who is paying the price of climate change, and who will increasingly pay the price as climate change intensifies? My University of Massachusetts colleague Jim Boyce has done outstanding work on this question for a long time. As a brief overview, he writes as follows:

> Rich countries burn more fossil fuels than do poor countries, generating more carbon dioxide emissions. And within any given country, richer people benefit more from the fossil-fuelled economy by virtue of the fact that

they consume more goods and services. Meanwhile it is poor countries and poor people who stand to bear the greatest costs of global warming. They are less able to invest in air conditioners, sea walls and other adaptations. They live closer to the edge . . . and the places that climate models show will be hit hardest by global warming—including drought-prone regions of sub-Saharan Africa and typhoon-vulnerable South and South East Asia—are home to some of the world's poorest people.[18]

It is also the case that shutting down the global oil, coal, and natural gas industries will inflict far greater costs proportionally on the workers and communities who now depend on the fossil fuel industry for their livelihood, relative to the owners—i.e. shareholders—of the private fossil fuel corporations. I have already discussed the situation for the corporations and their shareholders, but let me quickly review again the main points. Of course, the financial market valuation of the private fossil fuel companies will have to fall massively over the next two to three decades. The value of the oil, coal, and natural gas in the ground that is owned by these companies and that must not be burned if we are going to have any chance of stabilizing the climate has been estimated to be about $3 trillion. This is a formidable sum of money. But as I discussed earlier, this $3 trillion in corporate value tied to corporate ownership of unburnable oil, coal, and natural gas will decline on a fairly steady basis over, say, the next thirty years. That would amount to an average fall in value of the fossil fuel companies of $100 billion per year—still a lot of money, but nevertheless a mere pittance as a share of the total value of the global financial market as of

2019, which was around $317 trillion. This $100 billion annual decline would therefore amount to about 0.03 percent of the current value of the global financial market. Given these figures, it is certainly the case that any reasonably intelligent financial investor should recognize that the time has arrived to start getting out of fossil fuels and investing in other things.

Such a smooth transition path will not be an option for the workers and communities who are presently dependent on the fossil fuel industry for their livelihoods—that is, unless generous and effective just transition policies are established to support these workers and their communities. This is because their livelihoods, and the vibrancy of their communities, are 100 percent bound up with their jobs in the fossil fuel industry. When the fossil fuel enterprises are shut down in their communities, the workers' jobs will be lost, the value of their homes will plummet, and the tax revenues available to fund public schools, hospitals, public safety, street cleaning, public transportation, and parks will dry up.

This is exactly why, as I discuss in more detail later, just transition policies absolutely must be recognized as a front-and-center integral feature of any Green New Deal project worthy of the name, in all regions of the world. More generally, the global Green New Deal project must be focused on achieving net zero CO_2 emissions by 2050 via a path that is, in fact, right in front of us to be embraced—that is, a project that will also expand good job opportunities and raise living standards for working people and the poor throughout the world. The global Green New Deal can then also become the vehicle to defeat both the aggressively pro-corporate neoliberal variant of capitalism that has dominated global economics for two generations and the rise of neofascism that has emerged over the past decade as a warped pseudo-populist response to neoliberalism.

Following up on this issue of unequal impacts of climate change by countries, levels of development and class, certain leaders of the developing world, such as India's Prime Minister Narendra Modi, have been speaking about "climate justice." They are referring to the fact that their economies are not the ones who have created the climate crisis. Why then, in this view, should they have to bear the burden of sacrificing economic growth to fight climate change? What is your perspective on this "climate justice" argument, Noam?

NC: There is some justice in that position, to which we can add that the poor countries, who bear far less of a responsibility for the crisis, are its primary victims, India included. Nevertheless, when we consider the consequences—for these countries in particular—it would be suicidal for them to take this as a reason for delay in confronting the climate crisis. The right response, brought into the international agreements timidly and in far too limited ways, is for the rich countries to provide needed assistance in moving toward sustainable energy. And as I mentioned before, the Republican organization won't tolerate that.

Needed assistance could be provided in many ways, including very simple ones that could have considerable impact and would barely amount to a statistical error in national budgets. To take one example, much of India is becoming barely survivable because of more intense and frequent heat waves—reaching 50°C in Rajasthan in summer 2019. Those who can afford them are using highly inefficient and severely polluting air conditioners. That could be easily corrected. How much would it cost the rich countries to help people at least endure the fate that we have imposed on them, in our folly?

To be sure, this is a bare minimum. We can surely aspire to far more, even to the day when it is common understanding that the most vulnerable of both domestic and international society must be the prime objects of concern, and when institutions have undergone radical change so as to reflect and facilitate such common understanding. And just as we should be building the elements of a future society within the present one, following Bakunin's advice, so we should be seeking constantly to establish the sensibility of a more humane social order, hardly a novel idea and one that should always be a priority.

The earth's average temperature has already increased by about 1.1°C (2.0°F) since the 1880s. Just over roughly the past decade, this has caused record-breaking storms, forest fires, droughts, coral bleaching, heat waves, and floods around the world. Poor regions and communities are most vulnerable to these impacts, for the most basic reason that they have fewer resources to protect themselves. What major actions are needed now to protect people, communities, and the environment against these increasingly severe impacts that are already happening?

RP: Both parts of your question are worth emphasizing further. That is, let's first be clear that severe impacts of climate change are not just matters that our children, grand-children, and great-grandchildren may need to deal with, depending on which climate forecasting models end up being most accurate. The effects are happening right now. Thus, the 2019 *State of the Global Climate* report by the World Meteorological Organization (WMO) reports that "the physical signs and socio-economic impacts of climate change are accelerating as record greenhouse gas concentrations drive global temperatures towards increasingly danger-ous levels."[19]

And as you note, these climate-driven events will always inflict the most damage on low-income people and commu-nities. The reasons are straightforward. Dryland farmers rely on rain for irrigation and on forests to prevent floods. They are therefore most heavily impacted as the frequency and severity of droughts and floods increase. Poor people are then the ones who cannot afford to purchase the food they need when prices spike after a drought or flood. Low-income communities are also the ones with the least effective water drainage systems and with the fewest dykes and dams to ward off flooding.

The WMO's 2019 report offers dramatic evidence on these effects, highlighting Tropical Cyclone Idai, among other extreme weather events that occurred in 2018. Idai caused devastating floods, leading to more than 1,300 deaths in Mozambique, Zimbabwe, and Malawi, with thousands more missing. The study further described 281 floods occur-ring in 2018, impacting more than 35 million people, among them residents of the Indian state of Kerala, which suffered the heaviest rainfall and worst flooding in nearly a century. In 2018 we also saw a rise in world hunger resulting from weather extremes, after a prolonged period in which hunger and malnutrition had been declining. We should therefore expect the global figures on hunger, now at about 820 million people, or more than 10 percent of the world's population, to continue increasing along with the steady rise in the average global temperature.

If we go back just one year to 2017, Puerto Rico experienced Hurricanes Irma and Maria, with Maria having been the worst storm to have hit Puerto Rico in eighty years. According to various Puerto Rican government estimates, Irma and Maria caused nearly three thousand deaths, the loss of 80 percent of the island's crop value for the year, and as much as $90 billion in property damage, a figure roughly equal to 90 percent of Puerto Rico's GDP that year.[20] These climate change–induced disasters, in turn, only worsened the severe economic crisis that was already ongoing in Puerto Rico, inflicted by the punishing austerity policies that Wall Street and US policy makers had imposed on the island.

Comparable experiences will continue to become more frequent moving forward. It is therefore critical that a global Green New Deal project incorporate robust protections against climate change impacts. These protections should start with greatly expanding available storage facilities for food, seed, and fresh water, and ensuring that these structures are themselves strongly protected against climate events. It must then include water demand management infrastructure, including sea walls, dams, pumping capacity, permeable pavements, and abundant water-buffering vegetation. Existing buildings in vulnerable areas should be retrofitted to incorporate protective walls and green roofs to deal with both rainwater and heat. New buildings in vulnerable areas should be built with higher foundations or on stilts. Organic farming also provides important benefits in terms of climate protection, in addition to the advantages we have described above relative to corporate industrial farming. This is because organic farming is more effective than industrial agriculture at retaining the available water supply, using that water more efficiently, as well as mitigating soil erosion. Crop yields are also higher through organic farming under drought conditions and other forms of stress.

In addition to all these and additional forms of physical protections, people and communities need to have access to effective and affordable financial insurance against climate change damages. This will have to take the form of public insurance programs in almost all situations, since the costs of the relevant types of private insurance, such as against property or crop damages, are too high for virtually all the most vulnerable people. Public climate insurance programs could be implemented on an economy-wide basis, perhaps in conjunction with existing social insurance programs, such as social security or unemployment insurance. They could also be introduced on a smaller scale or more localized basis, as, perhaps, one innovative form of microfinance in poor communities.

Of course, any such adaptation measures will cost money. Various financial support programs are already operating throughout the world, both as specifically targeted climate adaptation measures, or as components of more general policies in the areas, for example, of infrastructure and housing development. But clearly, the current levels of such support are nowhere near adequate, especially, again, for low-income regions and communities. And here we have, on absolutely minimal grounds of basic decency, an obvious justification for the governments of high-income countries to be the major source of funding for these projects globally. They represent the countries that are responsible for having loaded up the atmosphere with greenhouse gases, while, to a greatly disproportionate extent, low-income people and communities now suffer the consequences of climate change already upon us.

While it is essential that all nations embark earnestly in the struggle to reduce carbon emissions, it is also obviously the case that the rich countries have vastly more resources for financing a Green New Deal than poor countries. This is true, even while the rich countries are the ones responsible for creating the crisis in the first place. In this context, it is imperative that we start talking about "climate finance." Bob, outline for us a realistic and feasible route for the investment project of a global Green New Deal.

RP: To begin with, it is important to jointly consider industrial policies and financial policies as an integrated framework for building a zero emissions global economy. Let's therefore consider these in turn.

Industrial Policies

There will need to be industrial policies that promote technical innovations and, even more broadly, adaptations of existing clean energy technologies. Each country's policy should be tailored to the specific conditions within it.

One major policy intervention that can facilitate the creation of a vibrant clean energy market is for governments themselves to become both large-scale investors in energy efficiency and purchasers of clean renewable energy. An important comparable historical experience was the development of the internet within the US military, beginning in the early 1950s. In the process of bringing the internet to commercial scale, the US military provided a guaranteed market for thirty-five years, which enabled the technology to incubate while private investors gradually developed effective commercialization strategies.[21]

But guaranteeing stable prices through private sector purchases of clean renewables is also critical here. Such policies are termed "feed-in tariffs." Specifically, these are contracts that require utility companies to purchase electricity from private renewable energy generators at prices fixed by long-term contracts. Feed-in tariffs were first implemented in the United States in the 1970s, and a number of state and local programs are currently operational in the United States today. However, the impact of feed-in tariffs has been much more significant outside the United States, especially in Germany, Italy, France, Spain, and Canada. The key factor in the success of these programs is straightforward: the prices for renewable energy were set to adequately reflect the costs of producing the energy along with ensuring a profit for the energy provider. This then encouraged private renewable energy investors by providing a stable, long-term market environment.

Another important set of policies are those that aim to directly reduce the consumption of oil, coal, and natural gas. These include carbon caps and carbon taxes, as I discussed earlier. In principle at least, a carbon cap establishes a firm limit on the allowable level of emissions for major polluting entities, such as utilities. Such measures will also raise the prices of oil, coal, and natural gas by limiting their supply. A carbon tax, on the other hand, will directly raise fossil fuel prices to consumers, and aim to reduce fossil fuel consumption through the resulting price signals. Either approach can be effective as long as the cap is strict enough, or the tax rate high enough, to significantly reduce fossil fuel consumption, and as long as exemptions are minimal to none. Raising the prices for fossil fuels will also, of course, create increased incentives for both energy efficiency and clean renewable investments, as well as a source of revenue to help finance these investments. I return to this point below.

However, significant problems are also associated with both approaches. Establishing a carbon cap or tax will have negative distributional consequences that will need to be addressed in the policy design. All else equal, increasing the price of fossil fuels would affect lower-income households more than affluent households, since gasoline, home-heating fuels, and electricity absorb a higher share of lower-income households' consumption. An effective solution to this problem is to rebate to lower-income households a significant share of the revenues generated either by the cap or the tax to offset the increased costs of fossil fuel energy.[22]

Renewable energy portfolio standards for utilities and energy efficiency standards for buildings and transportation vehicles are similar in their function as a carbon cap. That is, renewable portfolio standards set a minimum benchmark that utilities must achieve in generating electricity from renewable energy sources. Energy efficiency standards for automobiles set minimum miles-per-gallon levels (or comparable measures) that a given auto fleet must achieve to be in compliance with the law. Comparable efficiency standards can also be established for buildings in terms of acceptable levels of energy consumption for a given building size.

However, a major problem that has emerged with carbon caps as well as renewable and efficiency standards has been limitations in enforcement. As a major case in point, when these cap programs are combined with a carbon permit option—as in "cap-and-trade" policies—the enforcement of a hard cap becomes difficult to sustain or even monitor. The caps can be easily circumvented through taking advantage of complex trading requirements.[23] Within this context, the regulatory standards are simply ignored without anybody noticing. For example, New York State had set for itself a very modest renewable portfolio standard stipulating that, as of 2015, 29 percent of its electricity had to be generated from renewable sources. However, the state was only able to get to 21 percent renewable electricity by the deadline, despite the fact that 17 percent of its electricity had long been generated by hydropower plants that had been operating for decades. However, neither Governor Andrew Cuomo nor any other state official ever publicly acknowledged this failure, even while they have established substantially more ambitious clean energy targets moving forward. The point, therefore, is that all such regulations need to be strictly enforced.[24]

As I discussed earlier, one straightforward approach that is likely to get the attention of the relevant actors is to establish that any failure to meet a mandated renewable portfolio target should be punishable by jail time for the electrical utility's CEO.

Providing Cheap and Accessible Financing

In principle, it should not be especially challenging to solve this problem. To begin with, as of 2019, Credit Suisse estimates that the total value of global financial assets was $317 trillion. The $2.4 trillion that I am proposing to channel into clean energy investments beginning in 2021 amounts to 0.7 percent of this total financial asset pool.

Still, it is important to anchor the discussion in specific proposals. Therefore, for purposes of illustration, I propose four large-scale funding sources to support public investments in clean energy. Other approaches could also be viable. These four funding sources are: (1) a carbon tax, in which 75 percent of revenues are rebated back to the public but 25 percent are channeled into clean energy investment projects; (2) a transfer of funds out of military budgets from all countries, but primarily the US; (3) a Green Bond lending program, initiated by both the US Federal Reserve and the European Central Bank; and (4) the elimination of all existing fossil fuel subsidies and the channeling of 25 percent of those funds into clean energy investments. Strong cases can be made for each of these funding measures. But each proposal also has vulnerabilities, including around political feasibility. The most sensible approach is therefore to combine the measures into a single package that minimizes their respective weaknesses as standalone measures. The table in the appendix presents this set of combined proposals in summary form.

1. Carbon Tax with Rebates. As noted above, carbon taxes have the merit of shaping climate policy through two channels—they raise fossil fuel prices and thereby discourage consumption while also generating a new source of government revenue. At least part of the carbon tax revenue can then be channeled into supporting the clean energy investment project. But the carbon tax will hit low- and middle-income people disproportionately, since they spend a larger fraction of their income on electricity, transportation, and home-heating fuel. An equal-shares rebate, as proposed by James Boyce, is the simplest way to ensure that the full impact of the tax will be equalizing across all population cohorts.[25]

Consider, therefore, the following tax-and-rebate program. Focusing, again, on 2024, the first year of the full-scale investment program, we begin with a tax at a low rate of $20 per ton of carbon. Given current global CO_2 emissions levels, that would generate about $625 billion in revenue. Focusing on gasoline prices, a rule of thumb for estimating the impact of a carbon tax on retail prices is that every dollar in a carbon tax will add about one cent to the retail price per gallon of gasoline. Thus, starting the tax at $20 per ton will add about 20 cents to the price of a gallon of gasoline. As of 2020, the average retail price of gasoline globally was around $4, though the average price varies substantially by country, owing to differences in how gasoline is distributed and taxed. Nevertheless, as an illustrative average only, the carbon tax of $20 per ton would increase the average global retail price as of 2020 by 5 percent.

If we then use only 25 percent of this revenue to finance clean energy investments, that amounts to roughly $160 billion for investment projects. The 75 percent of the total revenue that is rebated to the public in equal shares would then amount to $465 billion. This amounts to about $60 for every person on the planet, or $240 for a family of four.[26]

2. Transferring Funds out of Military Budgets. Global military spending in 2018 was at $1.8 trillion.[27] The US military budget, at about $700 billion, accounted for nearly 40 percent of the global total. There are solid logical and ethical grounds for transferring a substantial share of each country's total military budget—if not most of it—to supporting climate stabilization, if we take seriously the idea that military spending is fundamentally aimed at achieving greater security for the citizens of each country. But to remain within the realm of political feasibility, let us assume that 6 percent of global military spending will transfer into supporting climate security. The 6 percent transfer of funds would apply to all countries on a proportional basis. The full amount of funds generated would be $100 billion.

3. Green Bond Funding by the Federal Reserve and European Central Bank. The response to the 2007–9 global financial crisis and subsequent Great Recession demonstrated that the Federal Reserve is able to supply basically unlimited bailout funds to private financial markets during crises. The extensive 2015 study, *The Cost of the Crisis*, by Better Markets concludes that the Federal Reserve committed approximately $12.2 trillion to stop the crash of the financial system, stabilize the economy, and encourage economic growth.[28] I would propose that the Fed supply $150 billion in Green Bond financing. This would amount to a minuscule 1.2 percent of its 2007–9 bailout operations during the crisis. The Fed's funding support could be injected into the global economy through straightforward channels. That is, various public entities, such as the World Bank, could issue long-term zero-interest-rate Green Bonds. The Fed would purchase these bonds. The various public entities issuing these bonds would then have the funds to pursue the full range of projects falling under the rubric of the global clean energy project.

As of this writing, this framework has not yet been introduced into policy discussions at the Federal Reserve. But green bonds are becoming a central area of focus at the European Central Bank. The *Financial Times* reported in December 2019 that the then recently installed ECB president Christine Lagarde was moving quickly on the matter: "Lagarde . . . is pushing to include climate change considerations in a review the central bank is due to hold into the way it conducts monetary policy. Because the central bank is by far the biggest influence on financial conditions in the market, it can make a significant difference to investment decisions that determine how Europe's climate transition goes."[29]

It is therefore reasonable to expect that the European Central Bank could also provide $150 billion in Green Bond financing, matching the Fed's contribution.

4. Eliminating Fossil Fuel Subsidies and Channeling 25 Percent of Funds to Clean Energy Investments. One recent estimate of direct fossil fuel subsidies to consumers—measured as the difference between supply and consumer prices for fossil fuel energy—is about $3 trillion globally as of 2015, or about 0.4 percent of global GDP.[30] Channeling these funds, in full, into supporting public clean energy investments would therefore more than pay for the $2.6 trillion estimate for total clean energy investments as of 2024. This $3 trillion would also represent more than double the amount necessary to cover a global public investment level of $1.3 trillion. However, such fossil fuel subsidies are largely used as a form of general support for all energy consumers. Lower- and middle-income households are therefore major beneficiaries of these subsidies, along with, of course, the fossil fuel corporations. Therefore, in terms of global income distribution, eliminating these subsidies altogether would likely have a significant regressive impact, comparable to establishing a carbon tax without an accompanying rebate program. As such, to continue to provide support for lower-income households, most of the funds that are now being channeled to these households through fossil fuel subsidies should be redirected into either supporting lower consumer prices for clean energy or to provide direct income transfers for lower-income households.

Given that we will have raised a total of $560 billion from the carbon tax ($160 billion), military spending transfers ($100 billion), and central bank Green Bond programs ($300 billion), we could then assume that 25 percent of the $3 trillion ($750 billion) received as fossil fuel subsidies would be transferred into the clean energy investment fund. With these funds, we will have reached the total $1.3 trillion in public investment funds necessary to attain the public portion of the combined public and private investment total ($2.6 trillion) as of 2024.

Channeling Financial Resources into Specific Investment Projects

Both general purpose development banks and special-purpose green development banks are already significantly engaged in financing clean energy investments. It will be crucial to build from these efforts to achieve the necessary level of private sector financing for clean energy investments.

The case of Germany is instructive, since it has been the most successful large advanced economy to date in developing its clean energy economy. KfW, the German publicly owned development bank, has been critical to this success. Stephany Griffith-Jones has examined KfW's impact on Germany's overall green transformation, including renewable energy as well as energy efficiency investments. She finds that KfW has underwritten roughly one-third of all financing for green investments in Germany. The bank has thus been instrumental in moving policy ideas into effective investment projects, with respect to both energy efficiency and clean renewables. KfW has also been highly active in financing green investment projects elsewhere in Europe and in developing countries. Griffith-Jones writes, "The combination of clear government policies and associated development bank targets has produced very positive results in green infrastructure in Germany, which can be replicated in emerging and developing countries."[31]

Griffith-Jones also describes the financing terms offered by KfW in all of their areas of active lending. These include a range of major subsidies for all their lending projects. With respect to financing clean energy investments in developing countries in particular, it is also critical that the benefits of these investments be shared fully by the society's least-advantaged groups. In a separate study, Griffith-Jones and her coauthors cite, as an example of an effective strategy, expanding access to electricity from clean renewable sources at low prices.[32] The authors stress that it is not realistic to expect clean energy investments to generate profits for private businesses at rates comparable to those in mature investment areas, including fossil fuel energy. The requirement that the financing terms for clean energy investments be affordable for borrowers reinforces the centrality of public investment banks with clear social criteria guiding their financing strategies.

Where Is the Money Coming from and Going To?

We need to be able to answer this question clearly to ensure that basic standards of fairness are built into the global Green New Deal. Let's start by reiterating three basic points:

1. Looking backward, the high-income countries, starting with the US but also including Canada, Western Europe, Europe, Japan and Australia, are primarily responsible for loading up the atmosphere with greenhouse gas emissions and causing climate change. They therefore should be primarily responsible for financing the global Green New Deal.

2. Moving from this historical perspective to the present, high-income people in all countries and regions have massively larger carbon footprints today than everyone else. As documented in a 2015 Oxfam study, the average carbon footprint of someone in the richest 10 percent of the global population is sixty times greater than of someone in the poorest 10 percent. The world's richest 1 percent is emitting as much as 175 times more than the poorest 10 percent.[33]

3. The up-front investment costs of a global Green New Deal are real and substantial, at around 2.5 percent of global GDP annually, amounting, as we have seen, to about $2.6

trillion in 2024. But these investments will pay for themselves over time, through dramatically raising energy efficiency levels and providing abundant clean renewable energy at average prices that are at parity with or lower today than those for fossil fuels and nuclear, and falling.

Within this overall framework, how well do my financing proposals measure up in terms of global fairness?

First, under my simple tax-and-rebate proposal, everyone on the planet receives a $60 rebate. For the average person in the US, this $60 will provide a tiny 0.1 percent boost to their income. But for the average person in, say, Kenya, this additional $60 will raise their income by roughly 6 percent. As a practical matter, individual governments will need to work out how to distribute these funds to their respective populations. In fact, there are many ways to implement a fair global carbon tax system.[34]

The impact of transferring 6 percent of all global military spending, provided on a proportional basis relative to current military budgets within each country, will also be strongly egalitarian. This is because, starting with the US, military spending levels of the high-income countries are much higher than those of middle- and low-income countries.

The Green Bond financing proposal will not take money out of anyone's pocket. Rather, it involves the world's two largest central banks in effect printing money as needed. This would be just as they did during the 2007–9 global financial crisis, except at a far more modest scale than the largesse that the central banks showered on Wall Street and the global financial elite to keep them afloat. To be clear, I am not suggesting that the US Fed or European Central

Bank should rely on this policy—what is technically known as "debt monetization"—on a routine basis. But we need to be equally clear that this is a fully legitimate option that the two major central banks have in their toolkit, and that this option should indeed be brought into action on a limited basis under crisis conditions. Note here that the funds will be generated by the central banks of high-income countries, but then distributed globally on an equitable basis, to underwrite the clean energy investment projects at scale in all regions of the globe.

Public investment banks in all regions, but especially in low-income countries, will then serve as primary conduits in moving specific investment projects forward. The public investment banks will be financing clean energy projects in both the public and private sector, along with mixed public-private projects. We cannot know what the best mix should be between public and private ownership with any specific project in any given country. There is no point in being dogmatic and pretending otherwise. But, in all situations, we do need to stick with the basic principle emphasized by Griffith-Jones and her coauthors: that with the private sector projects, it is not reasonable to allow private firms to profit at rates that they have gotten away with under forty years of neoliberalism. If private firms are happy to accept large public subsidies to support their clean energy investments, they then also need to be willing to accept limits on their profitability. Such regulatory principles are, for example, routine in the private US electric utility sector. They can be easily replicated elsewhere.

Many people are worrying that putting an end to reliance on fossil fuels would lead to massive job losses. But is it the case that the shift to clean energy resources would create new jobs and possibly encourage growth?

RP: The idea that building a green economy is a source of job creation should be intuitive, even though it is frequently portrayed in exactly the opposite manner—i.e., as a job killer. This is because building the green economy necessarily entails *building*—it means large-scale new investments to dramatically raise energy efficiency standards and equally dramatically expand the renewable energy supply. Spending money on virtually anything will create jobs. The only relevant question should then be *how many* jobs get created through building a green economy, and correspondingly, how many jobs will be lost through the contraction and eventual dissolution of the fossil fuel infrastructure.

In fact, countries at all levels of development will experience significant gains in job creation through clean energy investments relative to maintaining their existing fossil fuel infrastructure. Research that I have conducted with others has found this relationship to hold in Brazil, China, Germany, Greece, India, Indonesia, Puerto Rico, South Africa, South Korea, Spain, and the United States. For a given level of spending, the increases in job creation range from about 75 percent in Brazil to 350 percent in Indonesia. For India, as a specific example, Shouvik Chakraborty and I estimate that increasing clean energy investments by 2 percent of GDP every year for twenty years will generate an average net increase of about 13 million jobs per year. This would represent a gain of about 3 percent in overall jobs in the current Indian economy. This is also *after* factoring in job losses resulting from retrenchments in the country's fossil fuel industries.

Still, there is no guarantee that the jobs that clean energy investments generate will provide decent compensation to workers. Nor will these jobs necessarily deliver improved workplace conditions, stronger union representation, or reduced employment discrimination against women, minorities, or other underrepresented groups. But the fact that new investments will be occurring will create increased leverage for political mobilization across the board—for improving job quality, expanded union coverage, and more jobs for underrepresented groups.

At the same time, workers and communities throughout the world whose livelihoods depend on people consuming oil, coal, and natural gas will lose out in the clean energy transition. It is only a modest exaggeration to say that the fate of the planet depends on whether we can put in place just transition policies for these workers and communities that will be negatively impacted by the phase-out and shuttering of the fossil fuel industry. Just transition policies are certainly justified according to any standard of fairness. But they are also a matter of strategic politics. In the absence of such adjustment assistance programs operating at a major scale, the workers and communities facing retrenchment from the clean energy investment project will, predictably and understandably, fight to defend their communities and livelihoods. This in turn will create unacceptable delays in proceeding with effective climate stabilization policies.

Considering the US economy, Brian Callaci and I estimate that a rough high-end estimate for such a program is a relatively modest $600 million per year (that is, less than 0.2 percent of the 2018 US federal government budget).[35] This level of funding would provide strong support in two areas: (1) income, retraining, and relocation support for workers facing retrenchments; and (2) guaranteeing the pensions for workers in the affected industries. Comparable programs will of course need to be implemented in other country settings.

Another area that needs to be included in just transition discussions is reinvestment and general support for communities that are, at present, heavily dependent on the fossil fuel industry. These communities will face formidable challenges adapting to the fossil fuel industry's decline. One obvious set of projects would be to clean up and reclaim the land surrounding abandoned coal mines as well as oil and gas production sites. Another is to repurpose land. Germany's Ruhr Valley, which has been the traditional home for the country's coal, steel, and chemical industries, serves as a prominent case of successful repurposing. Since the 1990s, the region has advanced industrial policies to develop new clean energy industries. One important example in the region's repurposing initiative is RAG AG, a German coal-mining firm that is in the process of converting its Prosper-Haniel coal mine into a 200-megawatt pumped-storage hydroelectric reservoir. The repurposed facility will act "like a giant battery," with enough capacity to power more than 400,000 homes in North Rhine–Westphalia.[36]

An alternative to the proposed Green New Deal scheme for rescuing the planet from the catastrophic effects of global warming is the transition to a new economy beyond waste and continuous growth. This line of thinking has coalesced as a "degrowth" movement. Bob, in your view, is degrowth realistic or even desirable?

RP: I have been jousting with degrowth proponents for a few years. In my view, the basic considerations in the matter are straightforward. Yet, with a few exceptions, I don't seem to have made much progress in convincing most degrowth proponents. Let me try again here.

To begin with, I have lots of respect for most of the researchers and activists who advocate for degrowth. I share virtually all their values and concerns.[37] To be more specific, I agree that uncontrolled economic growth produces serious environmental damage along with increases in the supply of goods and services that households, businesses, and governments consume. I also agree that a significant share of what is produced and consumed in the current global capitalist economy is wasteful, especially much, if not most, of what high-income people throughout the world consume. It is also obvious that growth per se as an economic category makes no reference to the distribution of the costs and benefits of an expanding economy. As for Gross Domestic Product (GDP) as a statistical construct aiming to measure economic growth, there is no disputing that it fails to account for the production of environmental bads as well as consumer goods. GDP also does not account for unpaid labor, most of which is performed by women. GDP per capita also tells us nothing about the distribution of income or wealth.

Recognizing all these areas of agreement, it is still the case, in my view, that, on the specific issue of climate change, degrowth does not provide anything resembling a viable stabilization framework. Consider some very simple arithmetic. Following the IPCC, we know that global CO_2 emissions need to fall from their current level of 33 billion tons to zero within thirty years. Now assume that, following a degrowth agenda implemented as an emissions reduction program, global GDP contracts by 10 percent over the next thirty years. That would entail a reduction of global GDP four times larger than what we experienced over the 2007–9 financial crisis and Great Recession. In terms of CO_2 emissions, the net effect of this 10 percent GDP contraction, considered on its own, would be to push emissions down by precisely 10 percent—that is, from 33 to 30 billion tons. The global economy would still have not come close to bringing emissions down to zero, despite having manufactured the equivalent of a Great Depression in the effort to achieve this. Moreover, any global GDP contraction would result in huge job losses and declines in living standards for working people and the poor. Global unemployment rose by over 30 million during the Great Recession. I have not seen any degrowth advocate present a convincing argument as to how we could avoid a severe rise in mass unemployment if GDP were to fall twice as much as during 2007–9.

Clearly then, even under a degrowth scenario, the overwhelming factor pushing emissions down will not be a contraction of overall GDP but *massive growth* in energy efficiency and clean renewable energy investments (which, for accounting purposes, will contribute toward increasing GDP) along with similarly dramatic cuts in oil, coal, and natural gas production and consumption (which will register as reducing GDP). In other words, the global fossil industry will have to "degrow" to zero by 2050 while the clean energy industry massively expands.

These fundamental problems with degrowth are well illustrated by the case of Japan, which has been a slow-growing economy for a generation now, even while maintaining its average per capita income at a high level. Herman Daly, without question a major intellectual progenitor of the degrowth movement, himself describes Japan as being "halfway to becoming a steady-state economy already, whether they call it that or not."[38] Daly is referring to the fact that, between 1996 and 2015, GDP growth in Japan averaged an anemic 0.7 percent per year. This compares with an average growth rate of 4.8 percent per year for the thirty-year period 1966–1995. Nevertheless, as of 2018, Japan remained in the upper-income ranks among large economies, with average GDP per capita at about $40,000.

Despite the fact that Japan has been close to a no-growth economy for nearly twenty-five years, its CO_2 emissions remain among the highest in the world, at 8.8 tons per capita as of 2017. Moreover, Japan's per capita emissions have fallen only slightly since the mid-1990s. The reason is straightforward: as of 2017, 89 percent of Japan's total energy consumption still comes from burning oil, coal, and natural gas. Hydropower supplies about 2 percent of Japan's total energy consumption, while solar and wind supply another 2 percent.[39]

Thus, despite "being halfway to becoming a steady-state economy," Japan has made only modest progress, at best, in advancing a viable climate stabilization path. This is despite the fact that, at least in its official proclamations, the country has made commitments to rapidly expand its renewable energy sector and drive down emissions. Nevertheless, like all economies, large and small—whether they are growing rapidly or not at all—Japan needs to seriously follow through on its pledges to massively expand its clean renewable energy sector while its reliance on oil, coal, and natural gas "degrows" to zero.

Noam, how do you feel about the "degrowth" alternative to the climate change challenge?

NC: A shift to sustainable energy requires growth: construction and installation of solar panels and wind turbines, weatherization of homes, major infrastructure projects to create efficient mass transportation, and much else. Accordingly, we cannot simply say that "growth is bad." Sometimes, sometimes not. It depends on what kind of growth. We should of course all be in favor of the (very rapid) "degrowth" of energy industries, largely predatory financial institutions, the bloated and dangerous military establishment, and a lot more that we can list. We should be thinking about how to design a livable society—exactly as Bob has been doing. That will involve both growth and degrowth, raising many important questions. How it balances out depends on a wide range of particular choices and decisions.

Noam, if we assume that the wealthy nations will be contributing significant sums of money to developing countries so the latter can do their own part in staving off the effects of climate change, should we expect to see the rich nations making political and economic demands on the developing world? As such, could a new form of imperialism emerge in the relations between center and peripheral nations in the global capitalist economy? And, if so, should we also then expect to see political challenges against a global Green New Deal surfacing inside developing countries?

NC: I wish I felt more confident about the assumption. As already mentioned, the Republican organization is dead set against assisting poor countries, just as it is intent on savage punishment of Americans for the crime of not contributing enough to the coffers of the wealthy and deserving.[40] And there is not much vocal support even in less savage sectors of domestic and global society. It's perhaps worth mentioning that Americans have strange misconceptions about foreign aid. Polls show that they vastly overestimate its scale—but, when asked what they think it should be, give figures well above the actual minuscule level. That suggests that there may be prospects for organizing the public to support these critically important—and morally obligatory—measures. In general, what happens in this regard will depend on the consciousness, commitment, and strength of popular movements in the donor nations. Only that can fend off the kind of consequences you portray. All of these matters pose serious challenges for activists.

No doubt, if we reach the point where substantial and desperately needed aid is provided, policy makers in the rich countries will seek to impose conditions that would subordinate the recipients to their own priorities, much as IMF conditionalities do. But we are back to the same question.

Can popular movements reach a level of consciousness and scale sufficient to subvert such plans and ensure that genuine aid is provided to facilitate the difficult transition to sustainable energy policies? And much else.

Bob, proponents of the Green New Deal like you speak frequently of an economy of "full employment." What are the connections between advancing a global Green New Deal and supporting a full-employment economy?

RP: A commitment to full employment should be understood as fully consistent with and supportive of a global Green New Deal. There are several crucial interconnections here.

A full-employment economy is, most basically, one in which there is an abundance of decent jobs available for all people seeking work. Beginning with the individual's standpoint, whether you can get a job and, if so, whether the job offers decent pay and benefits, a clean and safe environment, and fair treatment for you and your coworkers matters a lot to almost everyone. An abundance of job opportunities is also crucial to an economy's overall health. As employment levels rise, so does total purchasing power in the economy, since people have more money in their pockets to spend. This means more buoyant markets, greater business opportunities for both small and large firms, and strong incentives for both public and private enterprises to increase their level of investment. This includes investments in building a green economy. An economy with an abundance of decent jobs will also promote both individual opportunity and equality, because this kind of economy offers everyone the chance to provide for themselves and their families. As such, full employment is also the single most effective policy for fostering social and economic equality.

Coming out of the 1930s Depression, the original New Deal, and World War II, creating full-employment conditions was the central focus of economic policy throughout the world. Of course, the level of commitment to this goal varied substantially by country and according to the political group in power. But it took the high inflation period of the 1970s and subsequent neoliberal revolution—marked most decisively by the elections of Margaret Thatcher as UK prime minister in 1979 and Ronald Reagan as US president in 1980—to supplant full employment as the centerpiece of economic policy in favor of a framework most amenable to Wall Street and global capitalists. This shift included macroeconomic policies focused on maintaining low inflation rather than full employment; reducing the public sector, including welfare state programs; eliminating or weakening pro-worker labor laws; eliminating barriers to international trade; and, of course, deregulating financial markets. It was the neoliberal commitment to financial deregulation in particular that proved to be the most direct cause of the 2007–9 Wall Street collapse and Great Recession.

The neoliberal revolution powerfully demonstrated a fundamental fact about full employment that was first recognized by Karl Marx, circa 1867, in *Capital*, vol. 1, in his famous chapter on what he called the "reserve army of labor." That is, capitalists will oppose full employment precisely because it raises workers' bargaining power relative to themselves. Wages will likely rise as the reserve army of labor is depleted and worker bargaining power correspondingly increases. Profit rates are then likely to be squeezed.

It is true that, over the decade following the Great Recession, we have seen the official unemployment rate fall sharply in the US, though less so in other high-income countries. US official unemployment as of early March 2020 was only 3.5 percent, as opposed to over 10 percent at the peak of the Great Recession.[41] But worker bargaining power has only nudged up slightly even during this low unemployment phase in the US. This is due to the combination of the reserve army of labor having expanded into a global pool of workers willing to accept jobs at much lower pay than the US standards, along with the US labor movement having been weakened by decades of persistent political attacks.

Full-employment policies will contribute powerfully to a viable global Green New Deal along several dimensions. As I noted above, it will create a supportive overall environment for the full range of investments needed for building a clean energy economy. It will also be critical in managing a just transition for workers and communities now dependent on the fossil fuel industry. This is because the single best form of protection for displaced workers in all countries is an economy that operates at full employment. In a full-employment economy, the challenges faced by displaced workers—regardless of the reasons for their having become displaced—are greatly diminished simply because they should be able to find another decent job without excessive difficulties. It also follows that, in a full-employment economy, the costs to taxpayers of providing reasonable levels of financial support for displaced workers would be greatly reduced. Also, in contrast to all variants of a neoliberal austerity agenda, a full-employment economy will increase the likelihood of attracting new investments, in clean energy and otherwise, into communities that have been hard hit by the loss of fossil fuel employment.

We have seen that, under a global Green New Deal, investments to build a clean energy economy will be a major engine of job creation. But we need to also recognize that the level of job creation generated by a global clean energy investment at about 2.5 percent of GDP will not be sufficient, on its own, to achieve and sustain full employment. These Green New Deal investments should succeed, on their own, to reduce official unemployment rates by 2 to 3 percent in most country settings. This is a substantial boost—to lower official unemployment, for example, from the March 2020 rate in Spain of about 14 percent to perhaps 11 percent, or from 29 to 26 percent in South Africa.[42] But in these countries and elsewhere, we will still need strong additional complementary policy interventions to overcome the ongoing neoliberal policy hegemony and advance a serious commitment to full employment.

The European Union (EU) has launched an ambitious European Green Deal plan to transform the 27-country bloc from a high- to a zero-emissions economy. Specifically, the plan, which involves overhauling every major aspect of the European economies, is to cut greenhouse gas emissions by 55 percent relative to 1990 levels by 2030 and to reach net zero emissions by 2050. European Commission president Ursula von der Leyen has called the project "Europe's man on the moon moment." Nevertheless, the plan has been sharply criticized by many environmental organizations and individual climate change activists alike, including the Swedish youth activist Greta Thunberg. Bob, what is your own assessment of the European Green Deal plan and how does it compare with your proposal for a global Green New Deal?

RP: In terms of its stated objectives, the European Green Deal is excellent. Its goals of achieving a 55 percent reduction in all greenhouse gas emissions by 2030 and to reach net zero emissions by 2050 are in full alignment with the IPCC's emission reduction target. The EU is therefore the only grouping of countries that has offered anything close to an official endorsement of the IPCC's targets. The European Green Deal also emphasizes just transition policies for workers and communities that are now dependent on the fossil fuel industries and, without such transition policies, will be hard hit as the fossil fuel firms are put out of business over the next twenty to thirty years.

That said, when one moves beyond the lofty rhetoric and commitments on paper, it becomes obvious that the European Green Deal is woefully inadequate. Certainly in politics, how much money someone is willing to spend to accomplish something is generally a moment-of-truth test of how truly committed they are to achieving their stated goals. By this standard, the European Green Deal isn't yet a serious project. To date, the project is budgeting 1 trillion euros over 2021–30 (1.09 trillion US dollars) for everything, including clean energy investments as well as just transition programs. This amounts to an average of about 100 billion euros per year in total spending. This is equal to only about 0.5 percent of total EU GDP per year over 2021–30. About half of the money would come from the EU budget, with the other half requiring additional funds from a combination of national governments and private investments.

In fact, as I have described above with respect to a global Green New Deal project, the amount of clean energy investment spending necessary to reach the net zero emissions goal by 2050 is about 2.5 percent of global GDP every year. Even if we allow that the EU countries could get by with a smaller investment spending commitment at about 2 percent of GDP per year, that would still imply an annual budget for 2021–2030 of roughly 400 billion euros per year—in other words, four times what the EU is proposing as its total Green Deal spending commitment. Indeed, the EU itself has stated that the budgetary requirements for meeting its 2030 emission reduction goals would have to be closer to 300 billion euros per year. It is as if the EU is trying to convey to the fossil fuel industry and its allies that the European Green Deal is not something they need to worry about excessively.

All that said, we still do not have the luxury of dismissing the European Green Deal outright. It is not meaningless that the EU has at least committed on paper to hitting the IPCC's emission reduction targets, as its "man on the moon moment." It is now the job of everyone who is truly committed to saving the planet to convert this rhetoric into a serious program. To make this happen, it will be crucial to continue making the case as loudly as possible that (1) Europe will not have to sacrifice jobs and living standards to hit what are now both the EU's and IPCC's emission reduction targets; and (2) building a clean energy infrastructure will save consumers money over time. This is because investments in energy efficiency, by definition, save money for consumers, and because generating energy from clean renewable sources is now already at cost parity or cheaper than energy from fossil fuels or nuclear power. If we can succeed in having these points sink in, we can also then make the legitimate case that the Green New Deal—in Europe as elsewhere—is also the egalitarian answer to still more decades of austerity, rising inequality, and diminished opportunities under neoliberalism.

Bob, in addition to your work on the global Green New Deal, you have also produced specific studies and proposals for many US states, as well as for the US as a whole, along with, among other countries, India, Spain, and Greece, and one colony, Puerto Rico. In considering all these individual studies for various states and countries, what is your assessment regarding how much of an impact the actions of any individual state or country can have in the race to save the planet?

RP: Of course, some individual countries matter a lot because their emissions levels, as a share of total global emissions, are outsized. Two countries, in particular, matter the most by this standard—China, whose CO_2 emissions represent 27 percent of the current global total, and the United States, which accounts for another 15 percent of current global emissions. So adding emissions from China and the US alone gets us to 42 percent of the global total. But we can also look at this same statistic from the opposite direction: even after combining the emissions levels for China and the US, we still haven't accounted for fully 58 percent of the global total. We can also push the individual country emissions figures a bit further and include all twenty-seven countries of the EU along with the United States and China. This adds another 10 percent to current total emissions, getting us to 52 percent of global emissions with China, the United States and the 27 EU countries combined. Therefore, if we only pay attention to China, the US and all EU countries, we still are neglecting the countries responsible for generating nearly half of current total global emissions.

The point is that every place does matter if we really are going to hit the target of net zero global emissions by no later than 2050. Zero emissions does mean *zero*, everywhere. If we let some small countries, or low-income countries, or a few low-population US states off the hook, then where do we draw the line and still get to the zero emissions goal?

Consider this example by way of illustration: CO_2 emissions right now in India are at 1.7 tons per person, which is one-eighth the average per capita emissions level for people living in the United States. This disparity reflects the fact that average incomes in India are only about 3 percent of those in the United States. But total emissions in India, at 2.2 billion tons, still account for nearly 7 percent of the global total, given that fully 18 percent of the world's population lives in India.

Now let's assume that the Indian economy grows at 3 percent per year through to 2050—a growth rate that is only about half the rate that India experienced over the previous thirty years—but this growth is powered by the exact same fossil-fuel-dominant energy infrastructure that fueled its growth over the previous three years. Under this scenario,

CO_2 emissions in India will have increased by nearly three-fold, to 5.5 billion tons, by 2050. In other words, the global economy will not come close to hitting the 2050 zero emissions target by allowing just this one exception for India, as a low-income country, to keep burning oil, coal, and natural gas. Now multiply the figures in the India example by the figures for all low-income countries in all regions of the world. If we allow for exceptions to the zero emissions standard based on countries' current average income levels, or population totals, or any other metric, we will simply foreclose the possibility of moving onto a climate stabilization project that has a chance of succeeding.

In recognizing this point, it is, once again, critical to also emphasize that a Green New Deal project for India—or for Kenya, Senegal, Greece, Spain, Colombia, Puerto Rico, and so on—is also a project that can raise mass living standards, expand job opportunities, and deliver clean air, soil, and water in all these countries. As such, the Green New Deal should be understood as the only viable framework for avoiding more decades under neoliberalism of severe economic crises and diminishing living standards as well as persistently rising average global temperatures.

Aside from the link between climate change and inequality, there's also the question of human migration. Indeed, it is feared that if the phenomenon of global warming continues unchecked the world will experience an unprecedented scale of human migration, mainly from the global South toward the West. But this nightmarish scenario may unfold even while we are in the process of adopting clean and renewable energy resources in order to contain the effects of climate change. In this context, what would be a humane but realistic immigration regime for the Western nations?

NC: This is a truly nightmarish scenario, and it's not in the future by any means. The UN currently estimates there are 65 million refugees fleeing from violence, persecution, and the impact of global warming. The poor countries are bearing the brunt of the massive problem of providing some minimal conditions for their survival: Kenya, Uganda, Bangladesh. The rich countries are in a turmoil because a small fraction of the refugee flow might reach them and defile their precious ethnic purity. Europe is funding Turkey to prevent people fleeing from Middle East horrors—for which the West has major responsibility—from reaching Europe. Europe is also providing what is called "development aid" to the world's poorest country, Niger; in reality this is the "development" of a security system to stop the flight of refugees from Africa, where Europe, it might be recalled, has a certain historical role in its plight.[43] Meanwhile thousands are dying in the Mediterranean, turned back by European force or the decision to just look the other way.

Much the same is true in North America. Trump's atrocious crimes should be familiar and I won't recount them. They raise earlier crimes to new levels of sadism in a manner peculiar to this administration. Militarization of the border took off under Clinton at the time when NAFTA was imposed, over the objections of most people in the affected states. It was well understood that NAFTA would destroy Mexican agriculture, provoking a flight of refugees. Campesinos may be efficient, but they cannot compete with highly subsidized US agribusiness, just as Mexican firms cannot compete with US conglomerates accorded "national treatment" under NAFTA (unlike Mexicans fleeing to the US).

But that's the least of it. There also should be no need to recount the horrifying US role in destroying Central America—though unfortunately the need to do so is real. The press currently features articles vividly portraying the miserable circumstances that impel flight from Guatemala, including many women fleeing from one of the world centers of femicide. The reports sometimes condemn the sadism of Trump's policy of driving refugees back to Guatemala so that they will not reach our borders while he cuts back the trickle of aid to Guatemala. They also deplore the cultural pathologies rooted in Guatemalan society that underlie the violence and crimes.[44]

But the reports somehow manage to miss the fact that Guatemala was overcoming these pathologies under the leadership of presidents Juan José Arévalo and Jacobo Arbenz in its decade of progressive reform, 1944–54—"years of spring in the country of eternal tyranny," in the words of a Guatemalan poet. Half a million people received land, including Indians who for the first time "were offered land, rather than being robbed of it," writes Latin American scholar Piero Gleijeses. "A new wind was stirring the Guatemalan countryside. The culture of fear was loosening its grip over the great masses of the Guatemalan population. In a not unreachable future, it might have faded away, a distant nightmare."[45]

But that was not to be. In 1954, something happened, something quite familiar in Latin America. The colossus of the North stepped in, overthrowing the government and re-imposing a harsh and murderous dictatorship. Since then Washington has regularly intervened to ensure the brutal rule of Guatemalan elites. Atrocities peaked under the murderous Reagan regime, when virtual genocide proceeded while the genial US president praised the worst monster, Ríos Montt, later condemned for genocide in Guatemala, as a man "totally dedicated to democracy" who was getting a "bum rap" from human rights organizations. And when Congress impeded the direct flow of arms to the mass murderers, the Reaganites turned to others to fill the gap, primarily the far-right Argentine dictatorship that Washington strongly supported, and Israel, always ready to employ its expertise in repression and its military products in service to imperial power, now providing the standard gear for the Guatemalan military.

All such matters are missing in the passionate and evocative accounts of the misery of the refugees fleeing from deplorable cultural practices and social disintegration, and from intensifying environmental degradation. Much the same is true in Europe where the picture is, if anything, even more grotesque.

Pope Francis is quite right to describe the "refugee crisis" as a moral crisis for the West.

So "what would be a humane but realistic immigration regime for the Western nations?"

For the most part, refugees are not fleeing *to* but fleeing *from*. They would much rather live at home. The first step, then, is to make that possible—a moral imperative, given our role in destroying their countries. The second step is to establish humane asylum procedures. Even these preliminary steps are hard to imagine right now: both the US and

Europe are vigorously committed to the opposite. But even if this bare minimum of human decency can be achieved, it will not begin to deal with the nightmare. All the more reason to devote every effort to rebuilding societies the West has destroyed and to preventing the environmental catastrophe that is already a significant factor in their flight, and is sure to become far more so in the not-distant future unless firm action is taken.

4

Political Mobilization to Save the Planet

How will climate change affect the global balance of power?

NC: It depends on the course followed by global warming—or global heating, as the *Guardian* more realistically calls it. If policies and practices continue to follow the present course, the question will be moot. Organized social life will collapse.

Suppose that sanity prevails and some viable social order remains. Then much depends on its nature. The steps that must be taken to save life on Earth from cataclysm may also induce significant changes in the nature of human society and popular consciousness. It could become more humane and just in the course of the cooperative effort and international solidarity that will be required to face this impending disaster, in which case the concept of the "global balance of power" might become obsolete, or at least significantly less brutal in its essence.

Let us however suppose that such a stage of civilization is not reached, while steps are taken to sustain organized human life in some manner. Then we can expect the global South to take another beating. Large parts may become barely livable: South Asia, the Middle East, much of Africa. And the rich will not escape. Australia is at severe risk, while a Trump-style criminal drives his country to disaster. China has severe ecological problems. Russia is highly vulnerable to climate change and, unlike China, is doing little about it.

By a cruel irony of history, the country that is in the lead in its dedication to destroy the planet would then probably be the least likely to suffer severe damage in the short term, and to maintain the position of global hegemony it has held, scarcely challenged, since it emerged triumphant in World War II—after a mobilization for war that consumed almost half of GDP, far more than what would be needed to achieve net zero carbon emissions within a few decades and to stave off utter catastrophe.

Painful though it is to speculate about the global balance of power if a decent level of civilization remains out of reach, it might not be too different in structure than in the post–World War II period, though "power" may be even more ugly than in the past.

RP: I have little to add on Noam's pessimistic scenarios, other than to underscore his first critical observation, that we are truly courting ecological disaster if, as Noam put it, "policies and practices continue to follow the present course." Indeed, Noam's observation is fully affirmed in the recent edition of the International Energy Agency's (IEA) flagship publication, the 2019 *World Energy Outlook*. This is the most extensive and authoritative mainstream publication of its kind in the world. According to this edition, the IEA forecasts that if the world proceeds along its present course, what the IEA calls its "Current Policies Scenario," global CO_2 emissions will not fall *at all* by 2040 from their current level of 33 billion tons, but rather will increase to 41 billion tons.

Still more alarming is the IEA's forecast under what it calls the "Stated Policies Scenario." This scenario aims to take account, as the IEA puts it, of "policies and measures that governments around the world have already put in place, as well as the effects of announced policies, as expressed in official targets and plans."[1] Therefore, among other considerations, the Stated Policies Scenario is meant to take full account of the agreements reached at the UN-sponsored 2015 Paris Climate Summit. Coming out of the conference, all 196 countries formally recognized the grave dangers posed by climate change and committed to substantially lowering their emissions. Nevertheless, the IEA estimates that, even under the Stated Policies Scenario, global CO_2 emissions will still not fall *at all* by 2040, but rather continue to rise to 36 billion tons.

In short, these IEA forecasts stand in breathtaking contrast to the IPCC's targets, as discussed earlier, of lowering global CO_2 emissions by 45 percent as of 2030—that is, to about 18 billion tons of emissions—and to reach net zero emissions by 2050. Thus, taking account of these figures, we cannot avoid the conclusion that Noam is not exaggerating in the least in describing where we are heading if "policies and practices continue to follow the present course."

But what if, somehow, we manage to move toward Noam's more optimistic scenario, that "sanity prevails and some viable social order remains." This scenario will necessarily produce major shifts in the global balance of power, in addition to the rather substantial benefit of enabling human life on Earth to continue more or less as we know it. Under this optimistic scenario, we will have succeeded in, among other accomplishments, putting the global fossil fuel industry out of business. This will completely upend every existing geopolitical configuration that has anything to do with oil, starting, of course, with the Middle East and moving out from there. These will include big macro impacts, at the level of big-power state and transnational capital machinations. But, as I discussed above, it will also open up opportunities for the flourishing of all kinds of small-scale energy-related initiatives all over the world, including public, private, and cooperative enterprises. Most importantly, this will include projects to deliver affordable energy to the nearly 1 billion people, mostly in rural regions of low-income countries, who still do not have access to electricity as of today.

The range of economic policy possibilities will also greatly expand for all countries that currently depend on importing fossil fuels to keep their economies functioning. These countries will no longer have to toe the neoliberal IMF line—that is, to maintain, as their top macroeconomic policy priority, success in global export markets along with holding back domestic spending. The IMF argues that such austerity measures are necessary to ensure that the energy-importing countries always have on hand enough money to purchase their all-important energy supplies. In a post–fossil fuel era, these countries can instead focus on building their respective clean energy infrastructures and expanding opportunities more generally within their domestic economies.

By the same token, the countries that now benefit from being oil exporters will clearly need to wean themselves off this economic model and advance more sustainable development paths. This will create some significant challenges in the short run, but these can be readily overcome after some initial adjustment periods. In fact, many energy-exporting countries now suffer from what is known as the "resource curse." That is, they organize their entire economy around the easy money that flows to them from selling energy. High government officials who sell favors to foreign oil companies have grown especially accustomed to the perquisites they receive from this model. As a result of this resource curse, it is generally true that economies that are energy exporters do not consistently perform better by standard economic indicators than the energy importers. For example, between 2010 and 2015, the six sub-Saharan African economies that were net energy exporters grew at only slightly faster average rates than the twenty-two countries that were net importers.

How, if at all, does the experience with the global coronavirus pandemic, and the response to it, help shed light on how we should think about climate change and the prospects for a global Green New Deal?

NC: At the time of writing, concern for the crisis is virtually all-consuming. That's understandable. It is severe and is severely disrupting lives. But it will pass, perhaps at horrendous cost, and there will be recovery. There won't be recovery from the melting of the Arctic ice sheets and other consequences of the dread advance of global warming.

Not everyone is ignoring the existential crisis that is advancing inexorably. The sociopaths who are dedicated to accelerating the disaster continue to pursue their efforts, relentlessly. As before, Trump and his courtiers continue to take pride in leading the race to destruction. As the US was becoming the epicenter of the pandemic, thanks in no small measure to their folly, the White House cabal released its budget proposals. As expected, they call for even deeper cuts in health-care support and environmental protection, in fact anything that might benefit the irrelevant population, in favor of beefing up the bloated military and building Trump's Great Wall. And to add an extra touch of sadism, "the budget promotes a fossil fuel 'energy boom' in the United States, including an increase in the production of natural gas and crude oil."[2]

Meanwhile, to drive another nail into the coffin that Trump and associates are preparing for the nation and the world, their corporate-run EPA weakened auto emission standards, thus enhancing environmental destruction and killing more people from pollution.

Also as expected, fossil fuel companies are lining up in the forefront of the appeals of the corporate sector to the nanny state, pleading once again for the generous public to rescue them from the consequences of their misdeeds.

In brief, the criminal classes are relentless in their pursuit of power and profit, whatever the human consequences. It would be disastrous if their efforts are not countered, indeed overwhelmed, by those who have some concern for "the survival of humanity." It is no time to mince words out of misplaced politeness. "The survival of humanity" is seriously at risk on our present course, to quote a leaked internal memo of JPMorgan Chase, America's largest bank, referring specifically to the bank's genocidal policy of funding fossil fuel production.[3]

One heartening feature of the present crisis is the rise in community organizations that have started mutual aid efforts. These could become centers for confronting the challenges of unprecedented severity that are already eroding the foundations of the social order. The courage of doctors and nurses, laboring under miserable conditions imposed by decades of socioeconomic lunacy, is a tribute to the resources of the human spirit. There are ways forward. The opportunities cannot be allowed to lapse.

RP: In addition to the fundamental consideration that Noam has emphasized, there are several other ways in which the climate crisis and the coronavirus pandemic intersect. To begin with, a prime underlying cause of the COVID-19 outbreak, as well as other recent epidemics, including Ebola, West Nile, and HIV, has been the destruction of animal habitats through deforestation and related human encroachments, as well as the disruption of the remaining habitats through the increasing frequency and severity of heat waves, droughts, and floods. As the science journalist Sonia Shah wrote in February 2020, habitat destruction increases the likelihood that wild species "will come into repeated intimate contact with the human settlements expanding into their newly fragmented habitats. It's this kind of repeated, intimate contact that allows the microbes that live in their bodies to cross over into ours, transforming benign animal microbes into deadly human pathogens."[4]

It is also likely that people who are exposed to dangerous levels of air pollution will have faced significantly more severe health consequences than those who have been breathing cleaner air. Aaron Bernstein of Harvard's Center for Climate, Health, and the Global Environment states that "air pollution is strongly associated with people's risk of getting pneumonia and other respiratory infections and with getting sicker when they do get pneumonia. A study done on SARS, a virus closely related to COVID-19, found that people who breathed dirtier air were about twice as likely to die from the infection."[5]

A separate point that was raised widely over the worst months of the COVID-19 pandemic was that the responses in the countries that handled the crisis relatively effectively, such as South Korea, Taiwan, and Singapore, demonstrated that governments are indeed capable of taking decisive and effective action in the face of a crisis. The total death tolls from COVID-19 in these countries were negligible, and a return to normal life occurred relatively soon after governments imposed initial lockdown conditions. The point is that similarly decisive interventions could be successful in dealing with the climate crisis where the political will is strong and the public sectors are relatively competent.

There are important elements of truth in such views, but we also should be careful to not push the point too far. For example, some commentators have argued that a silver lining outcome from the pandemic was that, because of the economic lockdown, fossil fuel consumption and CO_2 emissions plunged along with overall economic activity during the recession. While this is true, I do not see any positive lessons here with respect to advancing a viable emissions program that can get us to net zero emissions by 2050. In my view, the experience rather demonstrates why a degrowth approach to emissions reduction is unworkable. That is, emissions did indeed fall sharply because of the pandemic and recession. But that is only because incomes also collapsed and unemployment spiked over this same period. As such, the experience reinforces for me the conclusion I discussed earlier—that the Green New Deal offers the only effective climate stabilization path, since it is the only one that does not require a drastic contraction (or "degrowth") of jobs and incomes to drive down emissions.

That said, a truly positive development that came out of the pandemic and recession is that progressive activists throughout many parts of the world have fought to include Green New Deal investments in their respective countries' economic stimulus programs. It is critical to keep pushing these initiatives forward and making sure that they succeed.

In support of that end, it is important to pay serious attention to issues around how best to launch various components of Green New Deal projects. The aim here is to maximize both the short-term stimulus benefits and longer-term impacts of Green New Deal programs. I know the importance of such considerations from personal experience working on the green investment components of the 2009 Obama stimulus program, in which $90 billion of the $800 billion total had been allocated to clean energy investments in the US. The principles underlying these investment components were sound, but the people who worked on the program in its various stages, including myself, did not adequately calculate the time that would be realistically required to get many of the projects up and running. We knew then that it was critical to identify "shovel-ready" projects—that is, ones that could be implemented on a large scale quickly so they could provide an immediate economic boost. But relatively few green investment projects were truly shovel-ready at that time. One important reason for this was that the green energy industry was then a newly emerging enterprise. The backlog of significant new projects was therefore thin. It is only moderately less thin today in virtually all countries.

This means that people designing Green New Deal stimulus programs need to identify the subgroup of green investment projects that can realistically roll into action at scale within a matter of months. One good example that should be applicable within almost all country settings would be energy efficiency retrofits of all public and commercial buildings. This would entail improving insulation, sealing window frames and doors, switching over all lightbulbs to LEDs, and replacing aging heating and air conditioning systems with efficient ones, preferably, where possible, with heat pumps. Programs of this sort could generate large numbers of jobs quickly for secretaries, truck drivers, accountants, and climate engineers as well as for construction workers. They could also deliver significant energy savings, and thereby reduced emissions, quickly and at relatively low cost. Building off such truly shovel-ready projects, the rest of the clean energy investment program could then be ramped up and thus provide a strong foundation to economies as they move out of recession and on to a longer-term sustainable recovery path.

Eco-socialism is increasingly becoming a major tenet of the ideological repertoire of green parties in European countries and elsewhere, which may be the reason for their increasing appeal with voters and especially the youth. Is eco-socialism a cohesive enough political project to be taken seriously as an alternative for the future?

NC: Insofar as I understand eco-socialism—not in great depth—it overlaps very closely with other left socialist currents. I don't think we're at a stage where adopting a specific "political project" is very helpful. There are crucial issues that have to be addressed, right now. Our efforts should be informed by guidelines about the kind of future society that we would like to see come into being, and that can be constructed in part within the existing society in many ways, some already discussed. It's fine to stake out specific positions about the future in more or less detail, but for now these seem to me at best ways of sharpening ideas rather than platforms to latch on to.

A good argument can be made that inherent features of capitalism lead inexorably to ruin of the environment, and that ending capitalism must be a high priority of the environmental movement. There's one fundamental problem with this argument: time scales. Dismantling capitalism is impossible within the time frame necessary for taking urgent action, which requires a major national—indeed international—mobilization if severe crisis is to be averted.

Furthermore, the whole discussion is misleading. The two efforts—averting environmental disaster, dismantling capitalism in favor of a more free and just and democratic society—should and can proceed in parallel. And can proceed quite far with mass popular organization. A few examples were mentioned earlier, among them Tony Mazzocchi's efforts to forge a labor coalition that would not only challenge owner-management control of the workplace but also be in the forefront of the environmental movement, and the missed opportunity to socialize major sectors of US industry. There's no time to waste. The struggle must be, and can be, undertaken on all fronts.

Bob, in your view, can eco-socialism coexist with the Green New Deal project? And, if not, what type of a politico-ideological agenda might be needed to generate broad political participation in the struggle to create a green future?

RP: In my view, details of rhetoric and emphasis aside, eco-socialism and the Green New Deal are fundamentally the same project. Or to be more specific, in my view, the Green New Deal, as we have discussed the term throughout this book, offers the only path to climate stabilization that can also deliver an expansion of good job opportunities and rising mass living standards in all regions of the world. The Green New Deal therefore defines an explicit and viable alternative to austerity economics on a global scale. My coworkers and I have worked on exactly this issue—advancing the Green New Deal as an alternative to austerity economics—in different country settings over the past few years, including Spain, Puerto Rico, and Greece, as well as the United States. To state the case more generally: the Green New Deal is, in my view, the only approach to climate stabilization that is also capable of reversing the rise in inequality and thereby defeating both global neoliberalism and ascendant neofascism.

Beyond the Green New Deal, I truly don't know what exactly one would mean by "eco-socialism." Do we mean the overthrow of all private ownership of productive assets, and the takeover of them by public ownership? As Noam suggested, does anybody seriously think that this could happen within the time frame we have to stabilize the climate, that is, within no more than thirty years? And are we certain that eliminating all private ownership will be workable or desirable from a social justice standpoint—i.e. from the standpoint of advancing well-being for the global working class and poor? How do we deal with the fact that most of the world's energy assets are already publicly owned? How, more specifically, can we be certain that a transition to complete public ownership would itself deliver zero net emissions by 2050? To me, the overarching challenge is to try to understand the alternative pathways to most effectively build truly egalitarian, democratic, and ecologically sustainable societies, putting all labels aside, and being willing, as Marx himself insisted, to "ruthlessly criticize" all that exists, including all past experiences with Communism/Socialism and, for that matter, all authors, including Marx himself. Indeed, my favorite quote from Marx is "I am not a Marxist."

It is true that, in our discussions here, we have only briefly touched on other "planetary boundaries" besides the climate crisis, including air and water pollution as well as biodiversity losses. I understand that the eco-socialist movement does give substantial attention to these other critical environmental issues in addition to climate change. I fully share their concerns and welcome the focus they bring to these issues. We have concentrated here on the climate crisis for the simple reason that it is the matter of greatest urgency.

Europe's civil disobedience movement, led by Extinction Rebellion protesters as a strategy to tackle the climate crisis and create a just and sustainable world, is growing by leaps and bounds, especially among young people, but also seems to annoy many citizens and may even be alienating the general public. Noam, can you share with us your thoughts on the strategy of massive civil disobedience as a way to tackle the climate emergency?

NC: I was involved in civil disobedience for many years, during some periods intensely, and think it's a reasonable tactic—sometimes. It should not be adopted merely because one feels strongly about the issue and wants to show that to the world. That tactic can be proper, but it's not enough. It's necessary to consider the consequences. Is the action designed in a way that will encourage others to think, to be convinced, to join? Or is it more likely to antagonize, to irritate, and to cause people to support the very thing we are protesting? Tactical considerations are often denigrated—that's for small minds, not for a serious, principled guy like me. Quite the contrary. Tactical judgments have direct human consequences. They are a deeply principled concern. It's not enough to think, "I'm right, and if others can't see it, too bad for them." Such attitudes have often caused serious harm.

I'm not answering your question directly because I don't think there is a general answer. It depends on the circumstances, the nature of the planned action, the likely consequences as best we can ascertain them.

Bob, where do you stand on this question?

RP: I would just add that any and all tactics that have a chance of moving us closer to solving the climate crisis should be considered seriously. This includes civil disobedience. But we do also have to consider that if civil disobedience actions succeed, for example, in shutting down roads and public transportation systems on weekdays, this then means that people can't get to work, parents can't pick up their kids at daycare, and sick people can't make it to the doctor's office. Such consequences will reinforce the view that already exists widely—whether it is fair or unfair—that climate activists don't really care about the lives of ordinary people. Anything that strengthens this view among the general public is going to be politically disastrous.

As it is, this view already gets nurtured when climate activists do not show genuine commitment to just transition programs for the workers and communities that will be hurt through the necessary shutdown of the global fossil fuel industry. Such views are further strengthened when climate activists favor the imposition of carbon taxes without 100 percent rebates for most of the population, starting with lower-income people. These rebates compensate people for the cost-of-living increases they will face simply by driving their cars or using electricity in their homes. The Yellow Vest movement that emerged in 2018 in France to oppose the carbon tax proposal of the thoroughly tone-deaf President Emmanuel Macron is one obvious case in point here.

So yes, let's certainly include civil disobedience as one of the tactics we deploy when it becomes clear that the tactic will be truly effective. But by "effective," I mean that we are succeeding in advancing a Green New Deal project capable of delivering a zero emissions global economy by 2050.

As we have discussed at various points in this book, neoliberalism is still dominant, and even more dangerous neofascist social movements are on the rise. In this context, the prospects of energizing voters in order to demand fundamental levels of political mobilization to confront the climate crisis do not appear particularly promising. In fact, it seems that it is mainly the youth who are insisting that we address climate change with the level of urgency it demands. In that context, what do you think it would take to turn things around and elevate climate change to the very top of the public agenda worldwide? Noam, let's start with you.

NC: It has become almost de rigueur these days to cite Gramsci's observation, from Mussolini's prison, on how "the old is dying and the new cannot be born; in this interregnum, a great variety of morbid symptoms appear." And understandably. It's to the point.

Neoliberalism may remain the dominant elite mantra, but it is visibly tottering. Its impact on the general population has been harsh almost everywhere. By now in the US half the population has negative net worth, while 0.1 percent hold more than 20 percent of wealth—as much as the lowest 90 percent—and the tendencies toward obscene concentration of wealth are increasing along with its direct impact on the decline of functioning democracy and social welfare. In Europe the impact is in some ways worse, even if somewhat cushioned by the residue of social democracy. And morbid symptoms are everywhere: anger, resentment, increasing racism and xenophobia and hatred of scapegoats (immigrants, minorities, Muslims . . .); the rise of demagogues who stoke these fears and exploit the social pathologies that surface in times of confusion and despair; and in the international arena, the emergence of a reactionary international headed by the White House and incorporating such likable figures as Bolsonaro, MBS, al-Sisi, Netanyahu, Modi, Orbán, and the rest. But such morbid symptoms are countered by rising activism on climate change and many other fronts. The new has not yet been born, but it is emerging in many intricate ways and it is far from clear what form it will take.

Much is unpredictable, but there are a few things that we can say with confidence: unless the new that is taking shape confronts the twin imminent threats to survival—nuclear war and environmental catastrophe—and does so forcefully and soon, it won't matter much what else happens.

Bob, what are your own thoughts on the matter?

RP: I will start with another entirely apt aphorism from Antonio Gramsci: "Pessimism of the mind; optimism of the will." That is, if we take climate science seriously and then look at where the world is today, the odds of us moving the world onto a viable climate stabilization path—and specifically, of hitting the IPCC's stated target of net zero CO_2 emissions by 2050—are shaky at best. On the other hand, to invoke Margaret Thatcher's famous dictum, "there is no alternative" to doing everything possible to accomplishing exactly these goals.

With respect to "optimism of the will," we can point to the rapidly growing tide of climate activism that has started to deliver some major breakthroughs. Most emphatically, this includes the September 2019 global Climate Strike, led by the remarkable Swedish teenager Greta Thunberg. Estimates are that between 6 and 7.5 million people participated in various actions at 4,500 locations in 150 countries.

The Climate Strike is reflective of equally significant, if less dramatically visible, developments around the world. One case in point has been the successful movement in the western Mediterranean countries, including Spain, France, and Italy, to outlaw new oil and gas exploration and drilling, as well as to phase out existing projects. These are all very recent political breakthroughs, starting around 2016. Indeed, in the case of Spain, from 2010 to 2014, with the country then suffering from the aftershocks of the global financial crisis and Great Recession, government officials signed more than one hundred permits with oil companies to start new exploration and drilling projects throughout the country. But environmental activists joined forces with business owners in the tourism industry to mount a successful resistance against this idea of fossil fuel development as an economic recovery plan. The government's efforts to counter the effects of the economic crisis by opening up the country to oil exploration and drilling were "a bad dream," in the words of one municipal official from the Spanish island of Ibiza. "We luckily woke up," he said.[6]

Climate activism of this type at the grassroots level throughout Western Europe has also led the European Commission to officially establish its European Green Deal project. The overarching aim of the project is for the entire continent to achieve the IPCC's goal of net zero emissions by 2050. As of early 2020, both legislative bodies of the EU, the European Council and European Parliament, had voted to endorse the project. Of course, for legislative bodies to pass resolutions is the easy job. Whether the Europeans have the will to follow through on these commitments remains an open question.

Comparable developments are also gaining momentum in the United States, despite the buffoonish climate denialism of President Donald Trump. Thus, in June 2019, New York State passed the most ambitious set of climate targets in the country, including carbon-free electricity by 2040 and a net zero emissions economy by 2050. The New York initiative follows from similar if somewhat less ambitious measures to date, in California, Oregon, Washington, Colorado, New Mexico, and Maine.[7] One major factor in these US state-level developments is the increasing participation of the mainstream labor movement. Union members have assumed major leadership roles in some cases. The key here is that these state-level measures now need to incorporate substantial just transition programs for workers and communities whose livelihoods now depend on the fossil fuel industry. These people and communities are facing major hits to their living standards in the absence of generous just transition programs. By bringing just transition considerations to the forefront of the climate movement, the unions are building on the legacy of the visionary labor leader Tony Mazzocchi that Noam discussed earlier.

Climate movements remain at modest levels throughout most low- and middle-income countries. But there is a reasonable chance that this will change quickly, as activism is growing, along with similar types of coalitions among environmentalists, labor groups, and some sectors of business that we are seeing in the US and Western Europe. One reason people are mobilizing is that, as a result of air pollution, virtually all the major cities in low- and middle-income countries are becoming unlivable, including Delhi, Mumbai, Shanghai, Beijing, Lagos, Cairo, and Mexico City. Aman Sharma, a young Climate Strike activist in Delhi, spoke to this issue, telling *The Guardian* in September 2019, "We are out here to reclaim our right to live, our right to breathe and our right to exist, which is all being denied to us by an inefficient policy system that gives more deference to industrial and financial objectives rather than environmental standards."[8]

A critical factor in advancing this movement, in the developing countries and elsewhere, will be to demonstrate unambiguously how climate stabilization is fully consistent with expanding decent work opportunities, raising mass living standards, and fighting poverty in all regions of the world. This needs to be recognized as the core proposition undergirding the global Green New Deal. Advancing a viable global Green New Deal should therefore be understood as the means by which "optimism of the will" comes alive in defining the political economy of saving the planet.

Appendix

A Framework for Funding
the Global Green New Deal

Investment Level for 2024—Year 1 of Investment Cycle: $2.6 trillion in public and private investments, at 2.5 percent of GDP

Clean Energy Investment Areas

- *Clean Renewable Energy: $2.1 trillion*
  Wind, solar, geothermal, small-scale hydro, low-emissions bioenergy

- *Energy Efficiency: $500 billion*
  Buildings, transportation, industrial equipment, grid and battery storage upgrades

Public Sources of Funds: $1.3 trillion

- *Carbon Tax Revenues: $160 billion*
  25 percent of revenues from tax; 75 percent returned to consumers as rebate

- *Transfers from Military Budgets: $100 billion*
  6 percent of global military spending

- *Green Bond Purchases by Federal Reserve and European Central Bank: $300 billion*
  1.6 percent of Federal Reserve Wall Street bailout support during financial crisis

- *Transfers of 25 percent of Fossil Fuel Subsidies: $750 billion*
  Total fossil fuel subsidies: $3 trillion
  75 percent of funds reserved for lower clean energy prices or direct income transfers for lower-income households

Private Sources of Funds: $1.3 trillion

- **Policies for Incentivizing Private Investors**

  Government procurement

  Regulations
  Carbon caps and taxes
  Renewable energy portfolio standards for utilities
  Energy efficiency standards for buildings and transportation vehicles

  Investment Subsidies
  Feed-in tariffs
  Low-cost financing through development banks and green banks

Source: Robert Pollin, "An Industrial Policy Framework to Advance a Global Green New Deal," in Arkebe Oqubay, Christopher Cramer, Ha-Joon Chang, and Richard Kozul-Wright, eds., *The Oxford Handbook of Industrial Policy* (Oxford, UK: Oxford University Press, 2020).

Notes

1. The Nature of Climate Change

1 Here and throughout, all of C. J. Polychroniou's questions to Chomsky and Pollin appear in italics.

2 Julian Borger, "Doomsday Clock Stays at Two Minutes to Midnight as Crisis Now 'New Abnormal,'" *Guardian*, January 24, 2019.

3 Alexandra Bell and Anthony Wier, "Open Skies Treaty: A Quiet Legacy Under Threat," armscontrol.org, January/February 2019; Tim Fernholz, "What Is the Open Skies Treaty and Why Does Donald Trump Want It Canceled?," *Quartz*, October 29, 2019; Shervin Taheran and Daryl G. Kimball, "Bolton Declares New START Extension 'Unlikely,'" July/August 2019, armscontrol.org.

4 Theodore A. Postol, "Russia May Have Violated the INF Treaty. Here's How the United States Appears to Have Done the Same," thebulletin.org, February 14, 2019.

5 Thomas Edward Mann and Norman Jay Ornstein, "Finding the Common Good in an Era of Dysfunctional Governance," *Dædalus*, amacad.org, Spring 2013.

6 Bradley Peniston, "The US Just Launched a Long-Outlawed Missile. Welcome to the Post-INF World," defenseone.com, August 19, 2019.

7   Anthropocene Working Group, "Results of Binding Vote by AWG, Released 21st May 2019," quaternary.stratigraphy.org.

8   Andrew Glikson, "Global Heating and the Dilemma of Climate Scientists," abc.net.au, January 28, 2016.

9   Raymond Pierrehumbert, "There Is No Plan B for Dealing with the Climate Crisis," *Bulletin of the Atomic Scientists*, 75:5, 2019, 215–21.

10 Timothy M. Lenton, "Climate Tipping Points—Too Risky to Bet Against," *Nature*, 575:7784, 2019.

11 "The Sixth Annual Stephen Schneider Award: Naomi Oreskes and Steven Chu," recording, climateone.org, December 15, 2016.

12 Damian Carrington, " 'Extraordinary Thinning' of Ice Sheets Revealed Deep Inside Antarctica," *Guardian*, May 16, 2019.

13 Oded Carmeli, " 'The Sea Will Get as Hot as a Jacuzzi': What Life in Israel Will Be Like in 2100," haaretz.com, August 17, 2019.

14 Carmeli, " 'The Sea Will Get as Hot as a Jacuzzi.' "

15 Jeffrey Sachs, "Getting to a Carbon-Free Economy," *American Prospect*, December 5, 2019.

16 Sondre Båtstrand, "More than Markets: A Comparative Study of Nine Conservative Parties on Climate Change," *Politics and Policy*, 43:4, 2015.

17 "Pompeo Says God May Have Sent Trump to Save Israel from Iran," BBC.com, March 22, 2019.

18 John R. Bolton, "To Stop Iran's Bomb, Bomb Iran," *New York Times*, March 26, 2015.

19 Lisa Friedman, "Trump Rule Would Exclude Climate Change in Infrastructure Planning," *New York Times*, January 3, 2020.

20 Livia Albeck-Ripka, Jamie Tarabay, and Richard Pérez-Peña, " 'It's Going to Be a Blast Furnace': Australia Fires Intensify," *New York Times*, January 2, 2020; "Anthony Albanese Backs Australian Coal Exports ahead of Queensland Tour," sbs.com.au, September 12, 2019; Sarah Martin, "Australia Ranked Worst of 57 Countries on Climate Change Policy," *Guardian*, December 10, 2019.

21 Tal Axelrod, "Poll: Majority of Republicans Say Trump Better President than Lincoln," *The Hill*, November 30, 2019.

22 Jacob Mikanowski, "The Call of the Drums," *Harper's Magazine*, August 2019.

23 Intergovernmental Panel on Climate Change, ipcc.ch.

24 This was documented most recently in a 2019 article by Alexander Petersen, Emanuel Vincent, and Anthony Westerling, "Discrepancy in Scientific Authority and Media Visibility of Climate Change Scientists and Contrarians," *Nature Communications* 10:1, 2019, 1–14.

25 Gernot Wagner and Martin Weitzman, *Climate Shock: The Economic Consequences of a Hotter Planet* (Princeton, NJ: Princeton University Press, 2015), 74–5.

26 Wagner and Weitzman, *Climate Shock*, 55.

27 Economists' Statement on Carbon Dividends Organized by the Climate Leadership Council, econstatement.org.

28 Mark Lynas, "Six Steps to Hell," *Guardian*, April 23, 2007. The article is based on his 2007 book, *Six Degrees: Our Future on a Hotter Planet*.

29 IPCC, *Climate Change and Land: Summary for Policymakers*, 2019. The 25 percent figure is a summary of the full set of results reported on p. 10, section A3 of this study. See also the graphic presentation on p. 8.

30 ILO, *World Employment and Social Outlook 2018: Greening with Jobs*, Geneva, ilo.org, p. 45.

31 Noriko Hosonuma et al., "An Assessment of Deforestation and Forest Degradation Drivers in Developing Countries," *Environmental Research Letters* 7:4, 2012.

32 Rod Taylor and Charlotte Streck, "The Elusive Impact of the Deforestation-Free Supply Chain Movement," World Resources Institute, June 2018.

33 Stibniati Atmadja and Louis Verchot, "A Review of the State of Research, Policies and Strategies in Addressing Leakage from Reducing Emissions from Deforestation and Forest Degradation (REDD+)," *Mitigation and Adaptation Strategies for Global Change* 17:3, 2012.

34 "How Clean Is Your Air?," stateofglobalair.org.

35 James K. Boyce, *Economics for People and the Planet: Inequality in the Era of Climate Change* (London: Anthem Press, 2019), 59–60.

36 Boyce, *Economics for People and the Planet*, 67.

2. Capitalism and the Climate Crisis

1 Andrew Restuccia, "GOP to Attack Climate Pact at Home and Abroad," *Politico*, September 7, 2015.

2 Ben Geman, "Ohio Gov. Kasich Concerned by Climate Change, but Won't 'Apologize,' for Coal," *Hill*, May 2, 2012.

3 Christopher Leonard, *Kochland: The Secret History of Koch Industries and Corporate Power in America* (New York: Simon & Schuster, 2019), 394.

4 Christopher Leonard, "David Koch Was the Ultimate Climate Change Denier," *New York Times*, August 23, 2019.

5 Leonard, "David Koch"; " 'Kochland': How David Koch Helped Build an Empire to Shape US Politics and Thwart Climate Action," *Democracy Now!*, August 27, 2019.

6 Lisa Friedman, "Climate Could Be an Electoral Time Bomb, Republican Strategists Fear," *New York Times*, August 2, 2019; Pew Research Center, "Majorities See Government Efforts to Protect the Environment as Insufficient," pewresearch.org, May 14, 2018; Nadja Popovich, "Climate Change Rises as a Public Priority. But It's More Partisan Than Ever," *New York Times*, February 20, 2020.

7 Isaac Cohen, "The Caterpillar Labor Dispute and the UAW, 1991–1998," *Labor Studies Journal* 27:4, 2003.

8 Drew Desilver, "For Most US Workers, Real Wages Have Barely Budged in Decades," pewresearch.org, August 7, 2018.

9 Dwight Eisenhower, *Speech to the American Federation of Labor, New York City*, September 17, 1952, eisenhowerlibrary.gov.

10 Connor Kilpatrick, "Victory over the Sun," *Jacobin*, August 31, 2017; Derek Seldman, "What Happened to the Labor Party? An Interview with Mark Dudzic," *Jacobin*, October 11, 2015.

11 John Bellamy Foster, "Marx's Theory of Metabolic Rift: Classical Foundations for Environmental Sociology," *American Journal of Sociology* 105:2, 1999.

12 Paul Bairoch, *Economics and World History: Myths and Paradoxes* (Chicago: University of Chicago Press, 1995), 54.

13 Dana Nuccitelli, "Millions of Times Later, 97 Percent Climate Consensus Still Faces Denial," thebulletin.org, August 15, 2019.

14 Ben Elgin, "Chevron Dims the Lights on Green Power," *Bloomberg Businessweek*, May 29, 2014.

15 Michael Corkery, "A Giant Factory Rises to Make a Product Filling Up the World: Plastic," *New York Times*, August 12, 2019.

16 Graham Fahy, "Trump Refused Permission to Build Wall at Irish Seaside Golf Course," reuters.com, March 18, 2020.

17 Juliet Eilperin, Bardy Ennis, and Chris Mooney, "Trump Administration Sees a 7-Degree Rise in Global Temperatures by 2100," *Washington Post*, September 28, 2018.

18 Neil Barofsky, *Bailout: An Inside Account of How Washington Abandoned Main Street While Rescuing Wall Street* (New York: Free Press, 2012).

19 The Next System Project, thenextsystem.org.

20 Patrick Greenfield and Jonathan Watts, "JP Morgan Economists Warn Climate Crisis Is Threat to Human Race," *Guardian*, February 21, 2020.

21 Kimberly Kindy, "Jeff Bezos Commits $10 Billion to Fight Climate Change," *Washington Post*, February 17, 2020.

22 Andreas Malm, "The Origins of Fossil Capital: From Water to Steam in the British Cotton Industry," *Historical Materialism* 21:1, 2013, 35.

23 Malm, "The Origins of Fossil Capital," 33–4.

24 Karl Marx and Friedrich Engels, *The Communist Manifesto*, ed. L. M. Findlay (Peterborough, ON: Broadview Editions, 2004), 65.

25 Credit Suisse Research Institute, Global Wealth Report 2019, credit-suisse.com.

26 Noah Buhayar and Jim Polson, "Buffett Ready to Double $15 Billion Solar, Wind Bet," *Bloomberg Business*, June 10, 2014.

27 Andrew Bossie and J. W. Mason, *The Public Role in Economic Transformation: Lessons from World War II*, Roosevelt Institute, March 2020.

3. A Global Green New Deal

1 As recognized initially by William Stanley Jevons in 1865, raising energy efficiency can also generate "rebound effects"—i.e., the rise of energy consumption resulting from lower energy costs. But as briefly summarized by Pollin, such rebound effects are likely to be modest within the current context of a global project focused on reducing CO_2 emissions and stabilizing the climate. Among other factors, energy consumption levels in advanced economies are close to saturation points in the use of home appliances and lighting, and with auto transportation and heating/cooling, average rebound effects are likely in the range of 10 to 30 percent. Granted, average rebound effects are likely to be significantly larger in developing economies. Therefore, it is critical that all energy efficiency gains be accompanied by complementary policies (as discussed below), including setting a price on carbon emissions to discourage fossil fuel consumption. Also, expanding the supply of clean renewable energy will allow for higher levels of energy consumption without leading to increases in CO_2 emissions. Robert Pollin, *Greening the Global Economy* (Cambridge, MA: MIT Press, 2015), 40–5.

2 Mara Prentiss provides a valuable brief discussion on these issues in her article "The Technical Path to Zero Carbon," *American Prospect*, December 5, 2019.

3 See Alicia Valero et al., "Material Bottlenecks in the Future Development of Green Technologies," in *Renewable and Sustainable Energy Reviews* 93, 2018, 178–200.

4 This development is documented in Pieter van Exter et al., *Metal Demand for Renewable Electricity Generation in the Netherlands: Navigating a Complex Supply Chain*, Copper8, 2018.

5 Troy Vettese, "To Freeze the Thames," *New Left Review* 111, 2018, 66.

6 Prentiss, "The Technical Path to Zero Carbon."

7 Mark G. Lawrence et al., "Evaluating Climate Geoengineering Proposals in the Context of the Paris Agreement Temperature Goals," *Nature Communications* 9:1, 2018, 1–19.

8 "Global Effects of Mount Pinatubo," Earth Observatory, earthobservatorynasa.gov.

9 Lawrence et al., "Evaluating Climate Geoengineering Proposals," 13–14.

10 James Hansen et al., "Nuclear Power Paves the Only Viable Path Forward on Climate Change," *Guardian*, December 3, 2015.

11 International Energy Agency, *World Energy Outlook 2019*, iea.org, 91.

12 US Energy Information Administration, "Nuclear Explained," eia.gov.

13 Rachel Mealey, "TEPCO: Fukushima Nuclear Clean-Up, Compensation Costs Nearly Double Previous Estimate at $250 Billion," abc.net.au, December 16, 2016; "FAQs: Health Consequences of Fukushima Daiichi Nuclear Power Plant Accident in 2011," World Health Organization, who.int.

14 US Energy Information Administration, "Levelized Cost and Levelized Avoided Cost of New Generation Resources in the Annual Energy Outlook 2020," eia.gov, February 2020.

15 IEA, *World Energy Outlook 2019*, 91.

16 For a good source on overall emissions, see Hannah Ritchie and Max Roser, "CO_2 and Greenhouse Gas Emissions," ourworldindata.org, first published May 2017, revised in 2019.

17 David Roberts, "Wealthier People Produce More Carbon Pollution— Even the 'Green' Ones," vox.com, December 1, 2017.

18 James K. Boyce, *Economics for People and the Planet*, 7.

19 World Meteorological Organization, *State of the Global Climate 2019*, public.wmo.int.

20 "Quick Facts: Hurricane Maria's Effect on Puerto Rico," mercycorps.org; Associated Press, "Hurricane Death Toll in Puerto Rico More Than Doubles to 34, Governor Says," *Guardian*, October 3, 2017.

21 See Vernon W. Ruttan, *Is War Necessary for Economic Growth? Military Procurement and Technology Development* (Oxford University Press, 2006).

22 For an effective solution to the distributional problem, via "carbon dividends," see Boyce, *Economics for People and the Planet*. Azad and Chakraborty expand on the idea of an egalitarian carbon dividend program for the global economy: Rohit Azad and Shouvik Chakraborty, "The 'Right to Energy' and Carbon Tax: A Game Changer in India," *Ideas for India*, 2019.

23 See, for example, Preston Teeter and Jörgen Sandberg, "Constraining or Enabling Green Capability Development? How Policy Uncertainty Affects Organizational Responses to Flexible Environmental Regulations," *British Journal of Management* 28:4, 2017, 649–50.

24 See Robert Pollin, Heidi Garrett-Peltier, and Jeannette Wicks-Lim, *Clean Energy Investments for New York State: An Economic Framework for Promoting Climate Stabilization and Expanding Good Job Opportunities*, Political Economy Research Institute, 79–80.

25 James Boyce, *The Case for Carbon Dividends* (Cambridge, UK: Polity Press, 2019).

26 Azad and Chakraborty develop a more complex rebate structure that rewards residents of countries according to the emissions levels of each country. Azad and Chakraborty, "The 'Right to Energy.' "

27 "World Military Expenditure Grows to $1.8 Trillion in 2018," sipri.org, April 29, 2019.

28 Better Markets, *The Cost of the Crisis*, bettermarkets.com, July 2015. As we go to press, the Fed is now committing even larger sums of money to counteract the US and global economic collapse induced by the COVID-19 pandemic.

29 Martin Sandbu, "Lagarde's Green Push in Monetary Policy Would Be a Huge Step," *Financial Times*, December 2, 2019.

30 David Coady et al., "How Large Are Global Fossil Fuel Subsidies?," *World Development* 91, 2017. This study distinguishes between direct fossil fuel subsidies—what it terms "pre-tax" subsidies—and "post-tax" subsidies. They define post-tax subsidies as including those covering global warming damages, air pollution damages, and vehicle externalities, including congestion, accidents, and road damage, estimating that these subsidies amount to roughly 6 percent of global GDP. These are valuable calculations. But for the purposes of this discussion on financing, the standard, and much more narrowly defined measure of pre-tax subsidies, is more directly relevant.

31 Stephany Griffith-Jones, "National Development Banks and Sustainable Infrastructure; the Case of KfW," Global Economic Governance Initiative, 2016, 4. Griffith-Jones's conclusions are fully in line with those of other researchers. For example, in a 2013 report on the energy efficiency market, the IEA concluded that "Germany is a world leader in energy efficiency. Germany's state-owned development bank, KfW, plays a crucial role by providing loans and subsidies for investment in energy efficiency measures in buildings and industry, which have leveraged significant private funds." International Energy Agency, *Energy Efficiency Market Report 2013*, 149.

32 Stephen Spratt and Stephany Griffith-Jones with Jose Antonio Ocampo, "Mobilising Investment for Inclusive Green Growth in Low-Income Countries,"enterprise-development.org., May 2013. As one specific policy solution, Azad and Chakraborty propose a program for rapidly advancing the expansion of renewable energy supply in India. The proposal includes a carbon tax, with the revenues from the tax being channeled into clean renewable energy investments that will then supply free electricity to low-income communities, many of which still have no access to electricity. Rohit Azad and Shouvik Chakraborty, "Green Growth and the Right to Energy in India," *Energy Policy*, 2020.

33 "Extreme Carbon Inequality: Why the Paris Climate Deal Must Put the Poorest, Lowest Emitting and Most Vulnerable People First," *Oxfam Media Briefing*, December 2, 2015.

34 Chancel and Piketty describe various viable approaches: Lucas Chancel and Thomas Piketty, *Carbon and Inequality: From Kyoto to Paris*, Paris School of Economics, November 3, 2015. These are in addition to the approach described above by Azad and Chakraborty.

35 Robert Pollin and Brian Callaci, "The Economics of Just Transition: A Framework for Supporting Fossil Fuel–Dependent Workers and Communities in the United States," *Labor Studies Journal*, 44:2, 2019.

36 See, for example, Lorraine Chow, "Germany Converts Coal Mine into Giant Battery Storage for Surplus Solar and Wind Power," *EcoWatch*, March 20, 2017.

37 For example, matters of emphasis and rhetoric aside, there is almost nothing that I disagree with in Tim Jackson and Peter Victor, "Unraveling the Claims for (and against) Green Growth," *Science* 366:6468, 2019. Jackson and Victor are the two leading economists advocating for degrowth.

38 Herman Daly and Benjamin Kunkel, "Interview: Ecologies of Scale," *New Left Review* 109, 2018.

39 Sumiko Takeuchi, "Building toward Large-Scale Use of Renewable Energy in Japan," japantimes.co.jp, July 8, 2019.

40 Aimee Picchi, "Total Trump Food-Stamp Cuts Could Hit up to 5.3 Million Households," CBS News, December 10, 2019.

41 In fact, as we were copyediting this book in mid-April 2020, US unemployment had skyrocketed due to the COVID-19 pandemic and corresponding economic collapse. For the last two weeks of March and first two weeks of April, initial unemployment insurance claims reached 21.9 million, nearly 14 percent of the US labor force. These are unemployment figures unseen since the 1930s Great Depression, and even during the Great Depression, the rise in unemployment did not occur at anything like the lightning pace we are experiencing now. Nevertheless, at this moment, it is too soon to draw any reliable broad conclusions from this unprecedented experience.

42 Again, these figures for Spain and South Africa do not reflect the impact of the COVID-19 pandemic on employment conditions in these countries.

43 Rémi Carayol, "Agadez, City of Migrants," mondediplo.com, June 2019.

44 Azam Ahmed, "Women Are Fleeing Death at Home. The US Wants to Keep Them Out," *New York Times*, August 18, 2019; Kevin Sieff, "Trump Wants Border-Bound Asylum Seekers to Find Refuge in Guatemala Instead. Guatemala Isn't Ready," *Washington Post*, August 16, 2019.

45 Piero Gleijeses, *Politics and Culture in Guatemala* (Ann Arbor: University of Michigan Press, 1988).

4. Political Mobilization to Save the Planet

1 International Energy Agency, *World Energy Outlook 2019*, iea.org, 751.

2 "What's in President Trump's Fiscal 2021 Budget?," *New York Times*, February 10, 2020.

3 Greenfield and Watts, "JPMorgan Economists Warn Climate Crisis Is Threat to Human Race."

4 Sonia Shah, "Think Exotic Animals Are to Blame for the Coronavirus? Think Again," *Nation*, February 18, 2020.

5 "A Conversation on COVID-19 with Dr. Aaron Bernstein, Director of Harvard C-CHANGE," Center for Climate, Health, and the Global Environment, Harvard T. H. Chan School of Public Health, hsph .harvard.edu.

6 Eurydice Bersi, "The Fight to Keep the Mediterranean Free of Oil Drilling," *Nation*, March 24, 2020.

7 David Roberts, "New York Just Passed the Most Ambitious Climate Target in the Country," vox.com, July 22, 2019.

8 Sandra Laville and Jonathan Watts, "Across the Globe, Millions Join Biggest Climate Protest Ever," *Guardian*, September 20, 2019.

Index

Gramsci, Antonio, 69, 150, 151
Great Britain, 61
Great Depression, vii, 122
Great Recession, vii, 57, 65, 106, 117, 122, 123, 152
Greater Israel Project, 13
Greece: desertification of, 24; Green New Deal as alternative to austerity economics, 146; job creation in, 113; specific studies and proposals for, 127
Green Bond lending program, 104, 106–7, 108, 111–12, 156
Green New Deal: air pollution, elimination of, 37; capitalism, as project to save, 81–2; channeling financial resources, 108–10; cheap and accessible financing, providing, 104–8; climate stabilization path, as only effective way to, 143, 146; degrowth movement as alternative to, 116–20, 142–3; development banks, 108–10, 157; dietary changes as part of, 32–3, 86; disruptive technologies, use of, 84–6; eco-socialism, coexistence with, 146–7; economic and political feasibility of: generally, 73 *et seq.*; financial policies, 104–12; industrial policies, 100–3; economic inequalities, reduction of, 91–6; economic stimulus programs and, 143–4; financing of: generally, 110–12, 119–21; fairness in, 110–12; framework for funding Global Green Deal, 156–7; responsibility for, 110; full employment and, 121–4; geoengineering techniques, 84–6; industrial policies needed for, 100–3; inequality, as resulting in diminution of, 91–5; insurance policy, as equivalent of, 18; job loss and

creation (*see* job loss and creation); land use requirements, 77, 79–81; major engine of job creation, as, 124; only viable alternative to avoid effects of neoliberalism, 129; political challenges in developing countries resulting from, 120–1; priorities of, ix–x; protection for people from impacts of climate change, 96–100; protections against climate change impacts, 98–9; public investment: investment level for 2024, 156; need for, 68, 74–6, 107, 112, 143–4; raw materials needed for, 78–9; saving capitalism, as, 81–2; "shovel-ready" projects, 143–4; specific investment projects, 108–10, 112; technical challenges to, 77–8; technologies, use of, 82–6; transition policies, 94, 114–15; upfront investment costs, 110–11
green parties, 144–7
greenhouse gas emissions: generally, viii, 15–17, 24–5, 26, 29–31, 33, 61, 74; afforestation, effect of, 83; China, by, 127–8; degrowth agenda as reduction program, 117–18, 142–3; European Union, by, 128; IEA estimates through 2040, 137; India, by, 128–9; inequality, relationship with, 91–2; Japan, in, 118; United States by, 127–8; zero emissions target (*see* zero emissions)
Griffith-Jones, Stephany, 109, 112
gross domestic product (GDP): generally, 10, 22, 74, 116; degrowth policies, effect of, 117–18; effect of storms on Puerto Rico's, 98; European Green Deal, under, 125–6; flaws as statistical construct, 116; fossil fuel subsidies as percentage of, 107; Green New Deal costs as percentage of, 110–11;